21 世纪高等院校电气信息类系列教材

自动控制理论与应用实验指导

Experiment Handout for Automatic Control Theory and Application

戴亚平　编著

机 械 工 业 出 版 社

本书采用中英文对照的方式编写，通过理论与实例结合的方式，深入浅出地介绍了自动控制实验中的控制系统设计与实现方法，可供相关专业采用双语教学的高等院校使用。

全书共分10章，介绍了自动控制系统的建模及各种控制器的设计与实现，包括根轨迹校正、频率法校正、PID控制器设计、LQR控制器设计、模糊控制器设计、神经网络PID控制器设计、遗传算法控制器设计等实验系统。每个章节中都配有必要的实例，让读者能快速掌握控制器设计与实现方法。

本书面向高等院校的学习自动控制理论课程师生以及对自动控制理论有一定认识的工程技术人员，旨在帮助读者用较短的时间掌握控制器设计的技巧和方法，并提高读者的实践与动手能力。

图书在版编目（CIP）数据

自动控制理论与应用实验指导／戴亚平编著．—北京：机械工业出版社，2017.9

21世纪高等院校电气信息类系列教材

ISBN 978-7-111-59203-7

Ⅰ.①自…　Ⅱ.①戴…　Ⅲ.①自动控制理论-高等学校-教材　Ⅳ.①TP13

中国版本图书馆CIP数据核字（2018）第033415号

机械工业出版社（北京市百万庄大街22号　邮政编码100037）

策划编辑：尚　晨　　责任编辑：尚　晨

责任校对：张艳霞　　责任印制：常天培

唐山三艺印务有限公司印刷

2018年3月第1版·第1次印刷

184mm×260mm·9.5印张·220千字

0001-2500册

标准书号：ISBN 978-7-111-59203-7

定价：29.00元

前　言
PREFACE

“自动控制原理”是工程类学科的一门重要的专业基础学位课程，在电气工程、机械工程、航天工程等工科专业都会设置。该课程既包含数学与工程科学相结合的理论知识，又包含理论与工程设计实践紧密结合的实验内容。本书是为了配合自动控制理论实验课程的教学而撰写的。

编者经过多年的教学实践获得如下体会：针对“自动控制原理”的实验课程，如果有一套与多媒体教学、网络教学、实物实验教学相结合的新型设备，允许教师拿到课堂进行实验教学、学生课后自己进行实验训练、科研人员针对自己设计的控制算法进行反复实验，会极大地提高“自动控制原理”这门课程的教学效果。

为了实现“课堂教学实验、课后网络化实验、学习者自创实验”的工作目标，我们开发了一套“便携式单自由度机械臂”实验系统，该系统构成了自动控制原理的实验硬件基础；在这样的硬件基础上，又开发了一套软件系统与硬件系统配套使用。我们设计的这套实验系统适合初学者使用，也允许使用者自己开发相应的实验项目。此外本书针对高校常用的“球杆实验系统”和“倒立摆实验系统”也进行了智能控制实验设计方面的讲解。

我们所期望的实验教学不再是学生在规定的时间，在座位有限的实验室进行的自控实验课程。而是将实验教学做到与学生、教师随时随地在一起，使得实验过程也能够如同教科书一般时刻伴随着读者。基于这样的实验教学思想，我们编写了这本涵盖了经典控制理论中多个基本实验的教材，在此基础上也想抛砖引玉，能够为激发高校开发出更多适合于培养工程型人才的实验教学教程出一点力。

全书共分10章，针对数学建模、超前滞后校正、PID控制、智能控制等方面进行设计与实验，每章都有具体的实例与实验模板，目的在于让读者结合实例更加快捷地掌握控制系统的设计与实现方法。

本书主要面向工科院校学习自动控制原理课程师生，可作为大、中专院校相关专业的教学和参考用书，也可供有关工程技术人员和软件工程师参考。为了方便广大读者更加形象直观地使用此书，每一章都给出了实例操作过程的Matlab程序的源文件，供学习者参考使用。全文采用中英文对照的方式撰写，以适应当前大专院校双语教学的需求，也可以作为培养留学生的参考教材。

本书由北京理工大学自动化学院的戴亚平编著，刘向东，姜增如，许向阳等也为本书的编写提供了大量帮助。日本东京工科大学的大山恭弘教授与佘锦华教授于2006年向编者的研究室赠送了最初版本的单自由度机械臂系统，此后北京理工大学自动化学院的多届研究生参与了“便携式单自由度机械臂系统”的开发，目前已经是第4代产品。在此对他们的大力帮助与辛勤工作一并表示感谢。

由于时间仓促，加上编者水平有限，书中不足之处在所难免，欢迎读者联系1282892958@ qq. com批评指正，编者将不胜感激。

编者

CONTENTS

目　　录

Chapter 1 Introduction

第1章 概　　述

"Automatic Control Theory", "Modern Control Theory" and "Intelligent Contro" are very important, but hard to understand courses for automation major's students. There are many abstract concepts in these courses, such as the system's stability, controllability, convergence speed and anti - interference performance, etc. Currently, many universities have carried out some education innovation, trying to improve the students' interest through varieties of teaching methods. The goal is raising the quality of theory teaching. The main steps of education innovation include the following aspects:

(1) The Demo experiments are added in class teaching via multimedia equipment

Teachers modify the PPT courseware, insert video, photos, animations etc., by means of multimedia technology, to stimulate students' interest by visual, auditory, tactile, etc. The vivid teaching in classroom will attract students to attend the teaching activity, and improve the quality of classroom teaching effectively.

(2) Add computer - assisted instruction

By means of simulation software and other compute tools, a lot of deduction, graphics rendering, etc., will be replaced by only one or two "instractions" in program. It will save teaching period effectively. Through observation and analysis of simulation output

自动控制原理、现代控制理论以及智能控制对于自动化专业的学生而言，是非常重要的课程，也是比较难以理解的几门课。课程中一些概念比较抽象，如系统的稳定性、可控性、收敛速度和抗干扰能力等。目前，各个高校都进行了不同程度的教学改革，力图通过多样的教学方式提高学生的学习兴趣，提升理论教学的教学质量。各高校的教学改革主要从以下几个方面进行：

（1）课堂教学借助多媒体设备加入演示实验

教师通过修改教学课件，在课件中插入视频、图片、动画等，恰当地应用多媒体技术，从视觉、听觉、触觉等多方面提升课程的可理解程度，激发学生的学习兴趣，鼓励学生积极参与课堂知识的学习，从而提高课堂的教学质量。

（2）引入计算机辅助教学

利用计算机仿真软件等教学工具，可以把大量的计算推导、图形绘制等工作仅用一两条指令代替，从而有效地节省教学时间。通过对仿真数据的观察和对输出图形的分析，加深了学生对基础理论的认识和理解。

data and graphics, it will enhance the students' understanding to basic theory.

(3) Offer experimental curriculum to aid theory teaching

The using of multimedia and computer aided instruction often cannot satisfy the requirements of engineering courses, because these experiments do not combine the physical plants. In recent years, due to the demand of "Engineer Accreditation", "Major Degree Certification" in universities to, a large number of experiments course are carried out in undergraduate and postgraduate education for assisting the theory teaching, so that the major knowledge will link the practical application closely. These curriculums which combined with the practice should not only make students verify the theoretical knowledge, but also improve the ability of engineering application.

According to all above mentioned, multimedia based experiment, networked experiments and so on are the hotspots currently. Developing a suitable experimental course is an important task in universities' in and after class experiment teaching innovation. With years of teaching practice, if there is a new set of device combining multimedia teaching, online teaching, and physical experiment to aim at we find that the control curriculum, will greatly improve the teaching and studying quality. The device will be taken by teacher to classroom, will be exercised by students after class, will be used by researchers to test their own controller algorithm repeatedly.

(3) 开设实验课程辅助理论教学

多媒体与计算机仿真类的辅助教学，因为不能结合实物的对象进行实验，所以不能满足工程类课程的学习需求。许多高校因为有了类似于“工程师认证”、“专业学位认证”等教育教学需求，近年来在本科与研究生的教学中增设了大量的实验课课程用于辅助理论教学，将专业知识与实际应用紧密联系起来。这种能够结合实践的课程，可以使学生验证已经学过的理论知识，还可以在实验的过程中提高学生的工程应用能力。

综上所述，多媒体、网络化等多种类型的实验教学目前正处于发展的热点，如何开发出适合课堂与课后的实验教学课程，是目前高校实验教学改革的重要任务。经过多年的教学实践，我们发现，针对控制类的课程，如果有一套与多媒体教学、网络教学、实物实验教学相结合的新型设备，允许教师拿到课堂进行实验教学、学生课后自己进行实验训练、科研人员可以自己设计控制算法进行反复实验的实验设备，会极大地提高教学效果。

Based on the idea, we designed an experimental system, which is the set of motion control system with a simple model, and is not only suitable for beginners, but also allows users to develop their own experimental program. In order to achieve such teaching ideas, we prepared the experiments handout which covers all the experiments of classical control theory. We would like to test firstly, and believe there will be more universities to develop more suitable experimental teaching tutorial for the cultivation of engineering talents.

我们正是基于这样的思想，设计出一套实验系统，该系统具有简单的运动控制模型，适合初学者使用，也允许使用者自己开发相应的实验项目。我们还编写了这本涵盖了经典自动控制理论中所有实验的讲义，在此基础上也想抛砖引玉，激发高校开发出更多适合培养工程型人才的实验教学教程。

In order to complete the goal of "experiments in teaching class, networked experiment after class, creative experiments by user", we designed and implemented a set of experimental system named "Portable One Freedom Robot Arm" (POFR – Arm), which constitutes the hardware of automatic control experiment. In POF – Robot system, we use DC motor as actuator, the position of the motor moved as a controlled object. In order to observe the experimental results more conveniently, an "arm" is connected to the DC motor as position indicator. Single Chip Micro computer (SCM) is used as the logical control unit to control the POFR – Arm position.

为了达到“课堂教学实验、课后网络化实验、学习者自创实验”的工作目标，我们首先设计并实现了一套“便携式单自由度机械臂”实验系统，该系统构成了自动控制实验的硬件基础。在该系统中，使用直流减速电动机作为实验装置的执行机构，减速电动机的位置作为系统的被控对象。为了更直观地观察实验结果，在电动机的出轴上接一个机械手臂，作为电动机位置的指示标志。使用单片机作为逻辑控制单元，对机械臂进行位置控制。

System hardware structure of POFR – Arm is shown in Figure 1. 1.

机械臂硬件系统结构如图 1. 1 所示：

Taking the POFR – Arm system as controlled object, we designed corresponding experiments course for "Automatic Control Theory": Mathematical modeling experiments, the root locus method correction experiments, frequency method correction experiments, pole

将该套单自由度机械臂系统作为被控对象，本书设计了对应“自动控制原理”课程的实验教程内容：系统的数学建模实验，根轨迹法校正实验，频率法校正实验，极点配置实验，PID 控制器设计实验，线性二次型最优控制器设计实验

configuration experiments, PID controller design experiments, linear quadratic optimal controller design experiments, etc. After the training of above experiments, students have a clear acknowledge and understanding for classical control theory, both the principles and the applications.

等。通过以上实验的训练后，学生对经典控制理论中的主要理论方法与应用都会有明确的认识与理解。

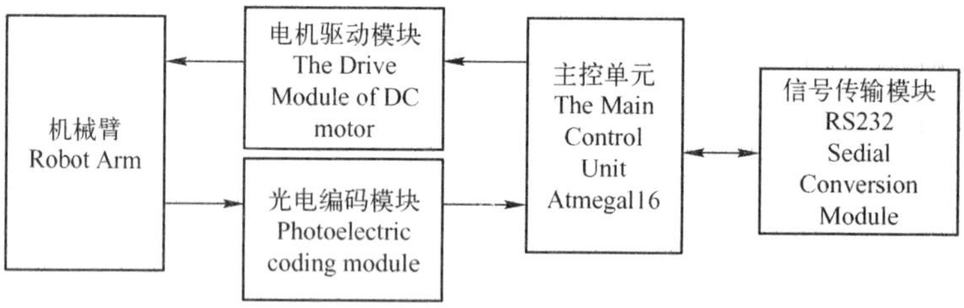

Figure 1. 1 System Hardware Structure of POFR – Arm

图 1. 1 便携式单自由度机械臂硬件系统结构图

Note that the experimental device in Figure 1. 1, The POFR – Arm can be replaced completely by other experimental device, and the experimental method designed can also be used for other controlled objects. Here we just take the POFR – Arm shown in Figure 1. 1 as an example to illustrate the use of the experimental teaching tutorials.

需要说明的是，图 1. 1 的实验装置完全可以被其他的实验装置所替代，本书所设计的实验方法同样可以用于其他被控对象的使用。这里只是将图 1. 1 所示的单自由度机械臂作为一个例子，来说明该套实验教学教程的使用。

Similarly, the designed system in Figure 1. 1 just only solved the experiment problem of being in the laboratory, limited by the time and space; it cannot meet the user's requirements of "anytime, anywhere" . So, we design a set of networked device shown in Figure 1. 2. With the appearance of this device, the experimental object in Figure 1. 1 was posted online, which allows students to use all types of internet tools in hands (computer, mobile phone, iPad, etc.) to do the experimental learning "anytime and anywhere" .

同样，图 1. 1 设计的系统仅仅是解决了实验者亲临现场的实验问题，受时间、地点的限制，并不能满足实验者"随时、随地"的要求。为此，我们设计了一套网络化装置如图 1. 2 所示。在增加了这套装置后，图 1. 1 的实验对象就被放到了网上，该套系统能够允许学生利用手中的各类上网工具（计算机、手机、iPad 等），进行"随时、随地"的实验性学习。

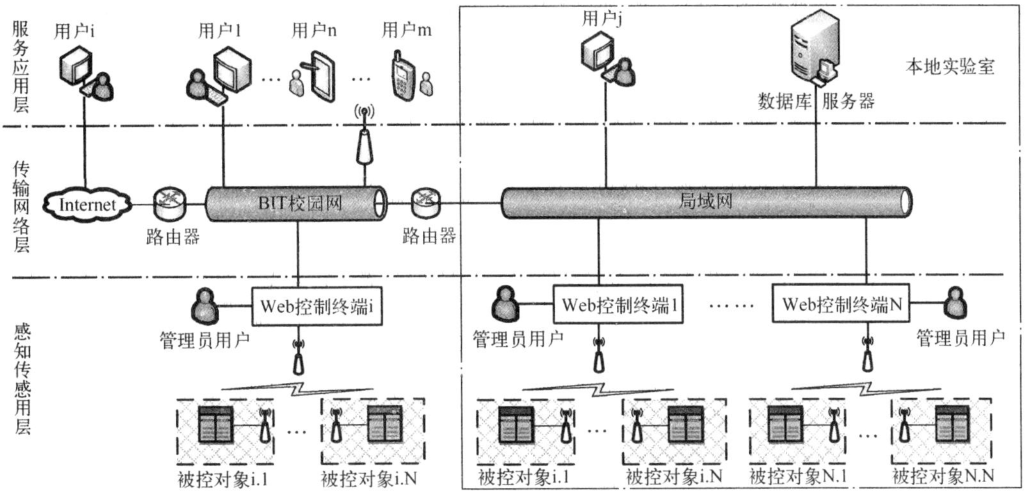

Figure 1. 2 Hardware Structure of Network Experimental System

图 1. 2 实验系统的硬件结构图

By the POFR – Arm, students can not only verify the theoretical knowledge taught by teachers in class, but also design their own experimental steps to verify control algorithms. The final results will be transmitted to PC or mobile terminals via the Internet, so that both the efficiency of the laboratory equipment and students study will be improved.

运用单自由度机械臂，学生们除了实验验证在课堂上老师教的理论知识，也可以自己设计实验步骤，来验证其他控制理论与控制算法的控制效果。最终的实验结果通过互联网传送回学生的个人电脑或手机终端上，实验设备的使用效率和学生的实践效率都大大提高了。

The laboratory equipment's connecting with "the Internet" is supported with more mature technology, which has a familiar name called "Internet of things". With the prosperity of intelligent mobile terminal market, we began to enter the mobile Internet era.

允许实验设备"上网"，有较为成熟的技术支持，这个技术有一个耳熟能详的名字称为"物联网"。随着移动智能终端市场的走向繁荣，我们开始进入移动互联网的时代。

As an extension of the mobile Internet, Internet of things has begun to walk into and change our life step by step. People began to connect traditional devices to the Internet to achieve the "objects" of the network, through the ubiquitous mobile network, then to achieve the remote monitoring and operation on a variety of devices anywhere and anytime. Remote

作为移动互联网的延伸，物联网一步步地走进并改变我们的生活。人们开始将传统的设备接入互联网，以实物联网也开始实现"物"的网络化。通过无处不在的移动互联网，实现对各种设备随时随地的远程监控与操作。（北京理工大学自动化学院）远程控制实验室，在有线网络化实验室的基础上，针对小型实

(Internet Based) Control System Laboratory (School of Automation, Beijing Institute of Technology), on the basis of wired network laboratories, designed and implemented a Embedded Web Control Terminal (EWC - T) based on wireless network for small equipment, and achieved a experimental network platform of "Wireless + Wired".

验设备设计并实现了基于无线网的嵌入式 Web 控制终端（缩写为 EWC - T），实现了一套"无线 + 有线"的网络化实验平台。

Based on Figure 1.1 and Figure 1.2, we designed new experimental teaching modes: 1) classroom experimental teaching mode; 2) homework experimental teaching mode; 3) multi - people interactive and discuss experimental teaching mode.

在图 1.1 和图 1.2 的设备基础上，我们设计出了新的实验教学模式：1）课堂实验教学模式；2）家庭作业实验教学模式；3）多人互动、讨论形式的实验教学模式。

These new modes are profoundly changing the way of talents training in engineering disciplines, and with the further application of Internet technology, there will be more experimental mode appear in college education.

这些新的实验模式都在深刻改变着工程学科的人才培养方式，并且随着互联网技术的进一步应用，可能还会有更多的实验教学模式在高等教育中涌现出来。

The purpose of this book is to provide students a new experimental platform, in which students can complete the task of teaching experiment originally done only in the laboratory. Now the experimental courses can be done by students and teachers anywhere anytime, which like your "carried - on" textbook in your learning career.

我们这本书的目的，就是提供给大家一个新的实验平台，在这个平台上可以完成原先只能在实验室里完成的实验教学任务。现在的实验课程可以做到与学生、教师随时随地在一起，如同你的教科书一般"时刻"伴随在你的学习过程中。

The main contents of this book are: The second chapter is the experimental modeling of second - order system; The third chapter is the root locus lead - lag compensation experiment; The fourth chapter introduces the frequency method lead - lag compensation experiment; The fifth chapter introduces the state space Pole assignment test; Chapter six

本书主要内容有：第 2 章讲述二阶系统的建模实验；第 3 章讲述根轨迹超前 - 滞后校正实验；第 4 章是频率法超前 - 滞后校正实验；第 5 章是状态空间极点配置实验；第 6 章是 PID 控制实验；第 7 章是线性二次型最优控制算法（LQR）控制实验；第 8 章讲述模糊控制系统的设计实验；第 9 章讲述基于 BP 神经元

introduces PID control experiments; Chapter seven introduces linear quadratic optimal control algorithm (LQR) control experiment; The eighth Chapter is the design of fuzzy control system experiment. The ninth chapter is the design of PID controller based on BP neuralnetwork method. The tenrh chapter is the experiment of controller design based on genetic algorithm. After learning the course of these experiments; students will have a deep understanding of control theory in engineering applications.

The born of this book benefits from the years of work by many post graduate students, and thanks to the support of 985 project funds. During the 985 construction equipment at the support of equipment department of Beijing Institute of Technology, we re - develop our purchased equipments, so that these experimental equipments could better serve teachers and students.

Moreover, through the development of these experimental equipments, students' practical and research abilities are improved effectively.

方法的 PID 控制器设计；第 10 章讲述基于遗传算法的控制器设计实验。通过这些实验课程的学习，学生可以加深对控制理论在工程应用的理解。

本书最终能够形成是得益于多届研究生的工作结果，并且得益于 985 工程经费的支持。在北京理工大学设备处支持下所进行的 985 工程建设中，我们对购置的实验器材进行了二次开发，使得这些实验器材能够更好地为教学服务。

通过开发这些实验设备，也培养了学生的动手与科研能力。

Chapter 2 Mathematical Modeling Experiments for Second - order System

第2章 二阶系统数学建模实验

Mathematical modeling is to use mathematical language to describe the behavior of a system. It is a process to get a formula from the real mechanics system, chemical system, aerospace system, etc.. It is the key step to design and analysis a control system. There are a lot of ways to model a system, and mostly will be done according to physical law. For example, we can build mathematical model according to the physical laws of the process as following: 1) Mechanical system principle; 2) Electrical system principle; 3) Fluid system principle; 4) Thermo dynamic system principle; 5) Chemical system principle; 6) Aerodynamics, etc.

系统的数学建模是对被控对象经过数学抽象后，建立公式类模型，即用数学语言来描述一个系统的行为。对于实际的机械、化工、航天等系统，获得一个切合实际的数学模型是分析系统、设计合理控制系统的关键。建立数学模型的方法多种多样，主要是根据被控对象内部存在的物理定律来建立数学模型，例如可以依据以下的系统原理进行数学建模：1）机械系统原理；2）电学原理；3）流体力学原理；4）热力学原理；5）化学原理；6）空气动力学原理等。

If the principal of system has been known, the modeling of dynamic system will be done as following steps:

在已经知道了系统的基本机理后，传统的方法可以依据下列步骤建立系统的数学模型：

1) Define the system and its components.

1）对系统以及各个部件进行定义；

2) Formulate the mathematical model after listing the necessary assumptions.

2）列写合理的假设条件之后建立数学公式；

3) Write differential equations describing the model.

3）用微分方程来描述模型；

4) Solve the equations for the desired output variables.

4）解微分方程以获得期望的输出；

5) Examine the solutions and the assumptions.

5）检验解与假设条件的正确性；

6) If necessary, reanalyze or redesign the system.

6）分析结果，如模型不合要求则重新设计系统。

2.1 The Principle of Modeling in POFR - Arm

2.1 便携式单自由度机械臂的数学建模原理

According to the definition of control system theory, a closed loop control system will be organized of plant, controller, adjustor, feedback equipment. The POFR - Arm is a motor control system, belongs to the type of motion control, and is shown as Figure 2.1. Its block diagram is shown in Figure 2.2.

The position of "arm" is the control variable, that will be represented as $\theta(t)$, and the θ_r is reference position (goal position), $\theta(t)$ is the current position of robot arm (true position).

根据自动控制理论对控制系统的定义，一个闭环控制系统需要包括被控对象、控制器、执行机构、反馈装置这几个部分，这样才能构成一个完整的闭环系统回路。单自由度便携式机械臂的控制系统是一个电动机控制系统，属于运动控制类，其实物如图 2.1 所示，系统的框图模型如图 2.2 所示。

机械臂的位置是我们的控制变量，也就是图中的 $\theta(t)$，其中 θ_r 是参考位置（目标位置），而 $\theta(t)$ 是机械臂的当前位置（实际位置）。

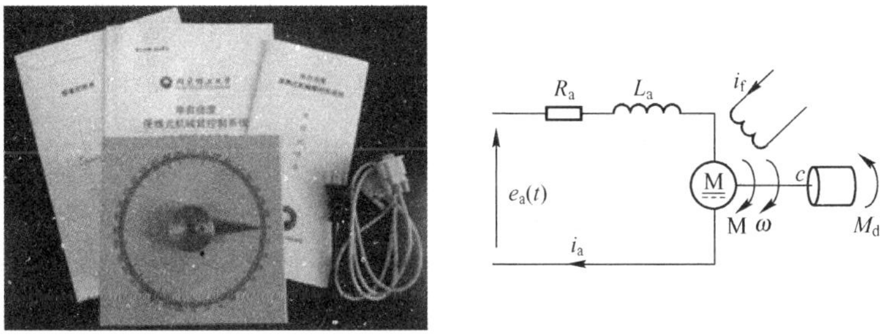

Figure 2.1 Diagram of Principle and Equipment for POFR - arm

图 2.1 单自由度便携式机械臂实物与原理图

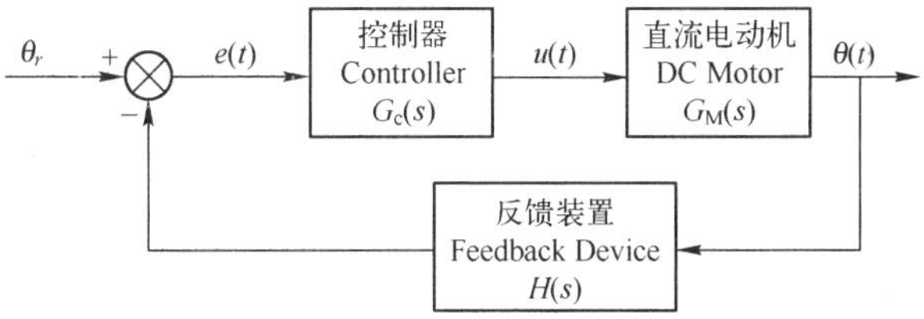

Figure 2.2 Diagram of Closed Loop System Block for POFR - Arm

图 2.2 单自由度便携式机械臂闭环控制系统模型

We will set up the mathematical model for DC motor firstly. As for there are no controller design, so we will assume the $G_c(s)=1$. Controller could be a kind of control algorithm, which is used as a compensator for whole system and will adjust the performance of dynamic and steady error, so how to design a controller is the most important experiment content, it will be discussed at next chapters.

我们首先要对直流电动机进行数学建模，最初的系统建立不包含控制器，可以假定 $G_c(s)=1$。控制器是控制系统的控制算法，它的作用就是校正系统的数学模型，调节系统的静、动态特性，是我们实验系统真正研究的内容，这一部分的设计在后续的章节中介绍。

The basic formula of DC motor comes from the theory of electromagnetic field which presented by Maxwell, the moment of DC motor is proportional to the product of magnetic flux φ and armature current i_a:

$$M=k_1\varphi i_a \tag{2.1}$$

Where k_1 is constant, M is the moment of DC motor.

直流电动机的基本方程基于麦克斯韦电磁理论，电动机产生的力矩正比于磁通量 φ 和电枢电流 i_a 的乘积：

$$M=k_1\varphi i_a \tag{2.1}$$

其中 k_1 为常数，M 为驱动电动机转动的力矩。

When the Motor in rotating, due to the relative motion of the armature winding and the magnetic field, the voltage e_b is generated in the armature winding, which is proportional to magnetic flux φ and angular velocity ω, shown in the formula (2.2):

$$e_b=k_2\varphi\omega \tag{2.2}$$

k_2 is constant. The voltage e_b and the armature winding voltage e_a has 180° phase angle difference, namely the direction of e_b is against the direction of voltage which makes the i_a flow, so e_b is called the counter electromotive force. If the excitation current i_f remains unchanged, changing the armature voltage e_a will control the DC motor.

在电动机转动的同时，由于电枢绕组和磁场的相对运动，在电枢绕组中产生正比于磁通量 φ 和角速度 ω 乘积的电压 e_b，得到式（2.2）：

$$e_b=k_2\varphi\omega \tag{2.2}$$

k_2 为常数，电压 e_b 和电枢绕组的外加电压 e_a 相位差 180°，即 e_b 的方向和使 i_a 流动的电压方向相反，因此称 e_b 为反电动势。若激磁电流 i_f 保持不变，用改变电枢电压 e_a 来控制直流电动机。

Assume that the total moment of inertia of the armature and the load of J, viscous friction coefficient is c, the armature circuit inductance and resistance for L_a and R_a, the load torque on the motor shaft is M_d, then on the basis of the voltage balance and moment balance principle, we have :

$$e_a - e_b = L_a \frac{\mathrm{d}i_a}{\mathrm{d}t} + R_a i_a \tag{2.3}$$

$$M = J\frac{\mathrm{d}\omega}{\mathrm{d}t} + c\omega + M_d \tag{2.4}$$

The magnetic flux φ is a function of the excitation current i_f in (2.1), and they keep in linear relationship when it is not in the saturation zone.

$$\varphi = k_3 i_f \tag{2.5}$$

k_3 is constant, so by the formula (2.1), (2.2) and (2.5), the following formulae can be obtained :

$$M = (k_1 k_3 i_f) i_a = k_m i_a \tag{2.6}$$

$$e_b = (k_2 k_3 i_f)\omega = k_b \omega \tag{2.7}$$

Define the electrical time constant $T_1 = L_a/R_a$, the mechanical time constant of $T_2 = J/c$, integrate the (2.6) and (2.7) into (2.3) and (2.4), we can obtain the following equation :

$$e_a - k_b\omega = R_a(T_1 D + 1) i_a \tag{2.8}$$

$$k_m i_a - M_d = c(T_2 D + 1)\omega \tag{2.9}$$

Because $\dot{\theta} = \omega$, connect equation (2.6) to (2.9) with a system block diagram, shown in Figure 2.3:

For permanent magnet DC motor, there is no excitation winding, the permanent magnet produces a fixed magnetic flux φ = constant. Let $M_d = 0$, and eliminate interme-

假定电枢和负载的总转动惯量为 J，粘性摩擦系数为 c，电枢电路的电感和电阻为 L_a 和 R_a，作用在电机轴上的负载力矩为 M_d，则依据电压平衡和力矩平衡原理有：

$$e_a - e_b = L_a \frac{\mathrm{d}i_a}{\mathrm{d}t} + R_a i_a \tag{2.3}$$

$$M = J\frac{\mathrm{d}\omega}{\mathrm{d}t} + c\omega + M_d \tag{2.4}$$

在式（2.1）中磁通量 φ 是激磁电流 i_f 的函数，在没有饱和的情况下呈线性关系。

$$\varphi = k_3 i_f \tag{2.5}$$

k_3 为常数，于是由式（2.1），式（2.2）以及式（2.5），可以得到下式：

$$M = (k_1 k_3 i_f) i_a = k_m i_a \tag{2.6}$$

$$e_b = (k_2 k_3 i_f)\omega = k_b \omega \tag{2.7}$$

定义电气时间常数为 $T_1 = L_a/R_a$，机械时间常数为 $T_2 = J/c$，将式（2.6）、式（2.7）式代入式（2.3）和式（2.4）式，则可以得到下式：

$$e_a - k_b\omega = R_a(T_1 D + 1) i_a \tag{2.8}$$

$$k_m i_a - M_d = c(T_2 D + 1)\omega \tag{2.9}$$

因为 $\dot{\theta} = \omega$，将式（2.6）至式（2.9）式用系统框图连接，如图 2.3 所示。

对于永磁式的直流电动机而言，没有激磁绕组，由永久磁铁产生固定的磁通 φ = 常量。令 $M_d = 0$，并消去中间变量 i_a 和 M_a，可得输入电压 $e_a(E_a(s))$ 和输

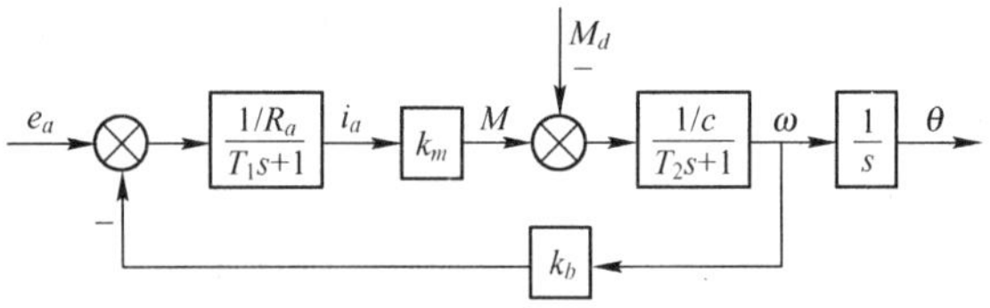

Figure 2. 3 Block diagram of DC motor controlled by armature

图 2. 3 电枢控制的直流电动机框图

diate variables i_a and M_a, we could have transfer function from input voltage $e_a E_a(s)$ and output angle speed ω ($\Omega(s)$):

$$\frac{\Omega(s)}{E_a(s)}=\frac{\frac{k_m}{R_a C}}{(T_1 s+1)(T_2 s+1)+\frac{k_b k_m}{R_a C}} \tag{2.10}$$

If let $e_a=0$, the transfer function should be determined by the load torque M_d with the output angle speed ω as following:

$$\frac{\Omega(s)}{M_d(s)}=\frac{(1/c)(T_1 s+1)}{(T_1 s+1)(T_2 s+1)+\frac{k_b k_m}{R_a C}} \tag{2.11}$$

(1) Put the parameters into (2. 11), the second - order system transfer function about ω is obtained. For position control, it becomes the third - order systems.

(2) If $T_1 \ll T_2$, ignore the impact of T_1, only T_2 is considered, the system transfer function will be written as follows:

$$\frac{\Omega(s)}{E_a(s)}=\frac{\frac{k_m}{R_a C}}{(T_2 s+1)+\frac{k_b k_m}{R_a C}} \tag{2.12}$$

$$\frac{\Omega(s)}{M_d(s)}=\frac{(1/c)}{(T_2 s+1)+\frac{k_b k_m}{R_a C}} \tag{2.13}$$

出转速ω($\Omega(s)$)之间的传递函数为:

$$\frac{\Omega(s)}{E_a(s)}=\frac{\frac{k_m}{R_a C}}{(T_1 s+1)(T_2 s+1)+\frac{k_b k_m}{R_a C}} \tag{2.10}$$

如果令 $e_a=0$，可以求出负载力矩 M_d 和输出转速 ω 之间的传递函数为:

$$\frac{\Omega(s)}{M_d(s)}=\frac{(1/c)(T_1 s+1)}{(T_1 s+1)(T_2 s+1)+\frac{k_b k_m}{R_a C}} \tag{2.11}$$

(1) 把参数带进式 (2. 11)，计算出二阶系统传递函数，得出关于 ω 的二阶系统。如果针对位置控制，则变成 3 阶系统。

(2) 如果 $T_1 \ll T_2$，则可以忽略 T_1 的影响，只保留 T_2，系统传递函数变成如下式所示的形式:

$$\frac{\Omega(s)}{E_a(s)}=\frac{\frac{k_m}{R_a C}}{(T_2 s+1)+\frac{k_b k_m}{R_a C}} \tag{2.12}$$

$$\frac{\Omega(s)}{M_d(s)}=\frac{(1/c)}{(T_2 s+1)+\frac{k_b k_m}{R_a C}} \tag{2.13}$$

Because $\Omega(s) = s\Theta(s)$, we could get the transfer function between input voltage and output rotation angle.

$$\frac{\Theta(s)}{M_d(s)} = \frac{(1/c)}{(T_2 s + 1) + \dfrac{k_b k_m}{R_a C}} \cdot \frac{1}{s} \tag{2.14}$$

由于 $\Omega(s) = s\Theta(s)$，故可以求出输入电压与输出转角之间的传递函数。

$$\frac{\Theta(s)}{M_d(s)} = \frac{(1/c)}{(T_2 s + 1) + \dfrac{k_b k_m}{R_a C}} \cdot \frac{1}{s} \tag{2.14}$$

It can be seen from equation (2.12), (2.13) and (2.14), that the relationship between the input voltage and the output rotation speed can be approximated as a first order system, and the transfer function of input voltage and output angle will be considered as the second order system.

由式（2.12）、式（2.13）以及式（2.14）可以看出，输入电压与输出转速之间可以近似为一阶系统，而输入电压与输出转角之间的传递函数则可以看作是二阶系统。

DC motor is the actuator of system, which controls the robotic arm rotation, torealize precise control of the robot arm's position. DC motor's parameters should be confirmed after it is produced and out of factory, so the mathematical model of the DC motor is determined by its electrical and mechanical characteristics properties. Feedback device is adopted of the optical encoder, which can detect the actual position of the robot arm, and will send the signal back to the controller.

直流电动机是系统的执行机构，它控制机械臂转动，实现机械臂的位置精确控制。直流电动机的自身特性在出厂之后即可视为固定不变，因此直流电动机的数学模型由自身的电气特性与机械特性决定。反馈装置由光电编码器实现，它可以检测到机械臂的实际位置，并将这一信号返回给控制器。

All the formulae from (2.1) to (2.14), are obtained according to the principle of DC motor. Assume that θ_r is a reference input which is given angular displacement (system input), the pointer is a plant of the control system, and pointer angular displacement θ_c is the controlled variables (system output), the measurement circuit is realized by a bridge potentiometer, it implements the tracking error$(\theta_r - \theta_c)$ of input and output of

以上的分析与式（2.1）~式（2.14），均是基于电机原理获得的，假定单自由度系统的参考输入为 θ_r，是给定角位移（系统的输入量）。单自由度机械臂的指针是被控对象，指针的角位移 θ_c 是被控变量（系统的输出量），测量电路是由桥式电位器来实现的，它实现系统输入量和输出量的跟踪偏差（$\theta_r - \theta_c$）并转换成电压信号 e_a，该信号经放大装置放大后驱动电动机，电动机则是整个系统的

the system and converted into a voltage signal e_a, the signal of e_a is amplified and then to drive the DC motor. The motor will be saying as the actuator of the system.

执行机构。

According to each component part structure, the system diagram can be obtained as figure 2.4.

根据各个环节结构可得整个系统的简化结构图如图 2.4 所示。

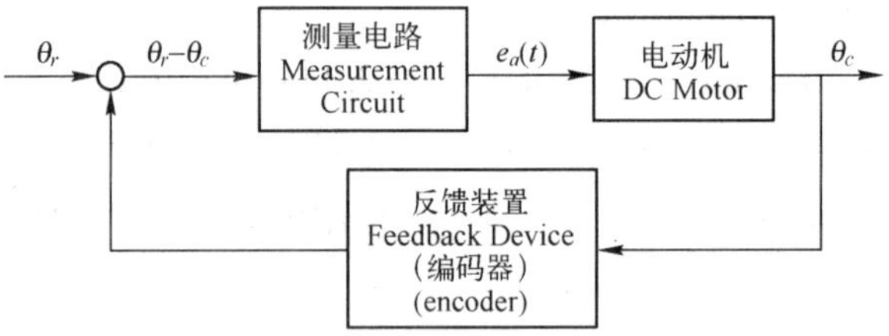

Figure 2.4 Simplified Block Diagram of POFR - Arm

图 2.4 单自由度机械臂简化框图

According to each component in POFR - Arm system, the structure diagram of the entire system can be obtained as show in figure 2.5. the parameters are: k_s is the parameters of bridge potentiometer, k_a is the gain of amplifier, T and k_d are constant of mechanical characteristics.

根据各个环节结构可得整个系统的结构图如图 2.5 所示。其中：k_s 为桥式电位器的传递系数，k_a 为放大增益；T，k_d 均为系统机械特性常数，可以认为是固定值。

So the closed feedback control system will be simplified as a classical second - order system, shown as figure 2.6. Where:

$$K=\frac{k_s k_a k_d}{c}$$

因此整个闭环系统可以近似简化为典型的二阶系统，如图 2.6 所示。其中：

$$K=\frac{k_s k_a k_d}{c}。$$

To facilitate the modeling, the transfer coefficient and amplifier gain of bridge potentiometer can be seen as a constant factor in the model of motor. Assume in the step response, the system stable speed is $\Omega(\infty)$, when the system rotation speed is $0.632\times\Omega(\infty)$, it will cost T seconds. According to the step response curve of motor position control, we can calculate the parameter K approximately.

为方便建模，把桥式电位器的传递系数和放大增益都可看成电动机模型中的常数系数。为了计算参数，假设系统在阶跃响应中，转速稳定时的速度为 $\Omega(\infty)$，当系统转速达到 $0.632\times\Omega(\infty)$ 时对应的时间就是 T。根据电动机位置控制的阶跃响应曲线，可以近似计算出参数 K。

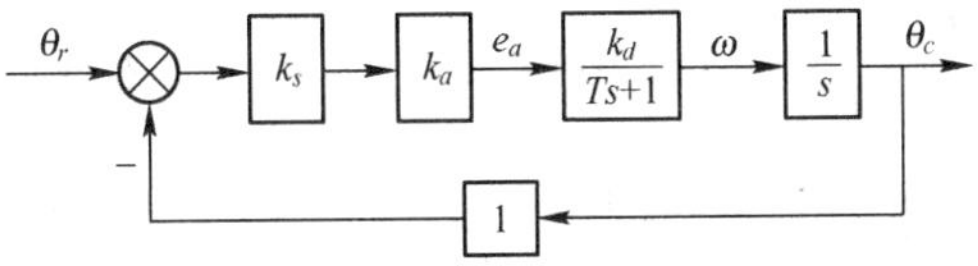

Figure 2. 5 Block Diagram of DC Motor

图 2. 5 电机控制框图

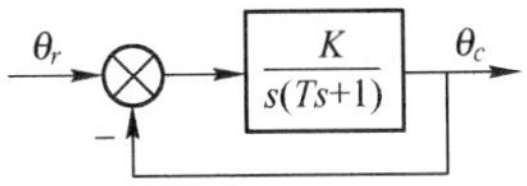

Figure 2. 6 Simplified Block Diagram of DC Motor

图 2. 6 电机控制框图的简化图

After several corrections and calculations, we will get the parameters of T and K as 0. 054 and 4. 587. Thus, the transfer function of motor input voltage and the output angle in DC motor can be obtained:

$$G(s)=\frac{4.587}{s(0.054s+1)} \tag{2.15}$$

The transfer function established $G(s)$ above is based on experiment with many times. Teachers can use the MATLAB program "least - squares parameter identification" at the end of the chapter to demonstrate the process of modeling for DC motor to students.

通过多次实验修正计算，最后得出系统 T 和 K 两个参数分别为 0. 054 和 4. 587。因此，直流电机输入电压与输出转角之间的传递函数是：

$$G(s)=\frac{4.587}{s(0.054s+1)} \tag{2.15}$$

以上传递函数的 $G(s)$ 确立，是在多次实验的基础上获得的。可以用本章附录的最小二乘参数辨识 MATLAB 程序，给学生展示电机数学模型的建过程。

2. 2 Mathematical Modeling Demonstration Experiments of POFR - Arm

2. 2 单自由度机械臂系统的数学建模实验

2. 2. 1 DC Motor Parameter Identification Experiment

2. 2. 1 直流电动机的参数辨识实验

The motor is a typical second - order system, and the transfer function block diagram is shown in Figure 2. 5.

Open loop transfer function:

电动机为典型的二阶系统，方框图及传递函数如图 2. 5 所示。

开环传递函数：

$$G(s)=\frac{K}{s(Ts+1)} \quad (2.16)$$

Where T, K are DC motor parameters.

Close loop transfer function:

$$H(s)=\frac{1}{(T/K)s^2+(1/K)s+1}=\frac{\omega_n^2}{s^2+2\xi\omega_n s+\omega_n^2} \quad (2.17)$$

In (2.17), $\omega_n=\sqrt{K/T}$, $\xi=\sqrt{1/4TK}$.

Formula (2.17) can be written to (2.18):

$$H(s)=\frac{s_1s_2}{(s+s_1)(s+s_2)} \quad (2.18)$$

Where

$$s_{1,2}=-\zeta\omega_n\pm\omega_n\sqrt{\zeta^2-1} \quad (2.19)$$

When the input is a step function, the inverse Laplace transform of the system is:

$$y(t)=L^{-1}\left[\frac{H(s)}{s}\right]$$

$$=1-\frac{s_2}{s_2-s_1}e^{-s_1t}+\frac{s_1}{s_2-s_1}e^{-s_2t} \quad (2.20)$$

Let$s_2=\alpha s_1$, then

$$1-y(t)=\frac{\alpha}{\alpha-1}e^{-s_1t}+\frac{1}{\alpha-1}e^{-\alpha s_1t}$$

$$=\frac{\alpha}{\alpha-1}e^{-s_1t}\left(1+\frac{1}{\alpha}e^{(-\alpha+1)s_1t}\right) \quad (2.21)$$

On both sides of the formula (2.21), put 10 as the base of the logarithm, there is,

$$\log(1-y(t))=\log\left(\frac{\alpha}{\alpha-1}\right)-s_1t\log e+\log\left(1+\frac{1}{\alpha}e^{(-\alpha+1)s_1t}\right) \quad (2.22)$$

$$G(s)=\frac{K}{s(Ts+1)} \quad (2.16)$$

式中，T，K 为电动机参数。

闭环传递函数：

$$H(s)=\frac{1}{(T/K)s^2+(1/K)s+1}=\frac{\omega_n^2}{s^2+2\xi\omega_n s+\omega_n^2} \quad (2.17)$$

在式（2.17）中，$\omega_n=\sqrt{K/T}$，$\xi=\sqrt{1/4TK}$。

可以将式（2.17）写为式（2.18）的形式：

$$H(s)=\frac{s_1s_2}{(s+s_1)(s+s_2)} \quad (2.18)$$

其中

$$s_{1,2}=-\zeta\omega_n\pm\omega_n\sqrt{\zeta^2-1} \quad (2.19)$$

当输入为单位阶跃函数时，对系统取拉普拉斯逆变换有：

$$y(t)=L^{-1}\left[\frac{H(s)}{s}\right]$$

$$=1-\frac{s_2}{s_2-s_1}e^{-s_1t}+\frac{s_1}{s_2-s_1}e^{-s_2t} \quad (2.20)$$

令 $s_2=\alpha s_1$，则有：

$$1-y(t)=\frac{\alpha}{\alpha-1}e^{-s_1t}+\frac{1}{\alpha-1}e^{-\alpha s_1t}$$

$$=\frac{\alpha}{\alpha-1}e^{-s_1t}\left(1+\frac{1}{\alpha}e^{(-\alpha+1)s_1t}\right) \quad (2.21)$$

对式（2.21）两边取以 10 为底的对数，有：

$$\log(1-y(t))=\log\left(\frac{\alpha}{\alpha-1}\right)-s_1t\log e+\log\left(1+\frac{1}{\alpha}e^{(-\alpha+1)s_1t}\right) \quad (2.22)$$

When $t \to \infty$, Let $\theta(t) = \log(1 - y(t))$ Formula (2.20) can be simplified as:

$$\theta(t) = at + b \qquad (2.23)$$

Where

$$a = -s_1 t\log e \approx -0.4343 s_1$$

$$b = \log\left(\frac{\alpha}{\alpha - 1}\right)$$

According to formula (2.21), and experimental data of step response shown in table 2.1, using the least squares method to identify the parameters A and B [see Appendix at the end of this chapter]. we can obtain:

$$a = -3.9896$$

$$b = 6.5241$$

Thus:

$$\alpha = \frac{1}{1 - 10^{-b}} = 1.000$$

$$s_1 = 9.188,\ s_2 = 9.188$$

$$\omega = \sqrt{s_1 s_2}$$

$$\zeta = \frac{s_1 s_2}{\sqrt{s_1 s_2}}$$

So we obtain the parameter of motor: $T = 0.054$, $K = 4.587$

So, the open loop system transfer function is:

$$G(s) = \frac{4.587}{s(0.054s + 1)} \qquad (2.24)$$

Also, the (2.24) will be written as following:

$$G(s) = \frac{84.42}{s(s + 18.376)}$$

当 $t \to \infty$ 时，令 $\theta(t) = \log(1 - y(t))$，式(2.22)可以化简为：

$$\theta(t) = at + b, \qquad (2.23)$$

其中：

$$a = -s_1 t\log e \approx -0.4343 s_1;$$

$$b = \log\left(\frac{\alpha}{\alpha - 1}\right)$$

对于式(2.21)，以及根据实验获得的表2.1所示的阶跃响应数据，采用最小二乘法辨识参数 a 和 b [见本章附录]，可以得到：

$$a = -3.9896$$

$$b = 6.5241$$

由此可得：

$$\alpha = \frac{1}{1 - 10^{-b}} = 1.000$$

$$s_1 = 9.188,\ s_2 = 9.188$$

$$\omega = \sqrt{s_1 s_2}$$

$$\zeta = \frac{s_1 s_2}{\sqrt{s_1 s_2}}$$

因此可以得到电动机的参数为 $T = 0.054$，$K = 4.587$。

于是得到系统的开环传递函数为：

$$G(s) = \frac{4.587}{s(0.054s + 1)} \qquad (2.24)$$

或者写为：

$$G(s) = \frac{84.42}{s(s + 18.376)}$$

Table 2.1 Step Response Value of DC Motor System

表 2.1 电机系统阶跃响应实验值

t/s	0.036	0.196	0.286	0.375	0.464	0.571	0.714	0.768
$y(t)$/rad	0	60	120	180	240	300	360	382

2.2.2 Experiment on Motor Step Response

(1) The purpose of Experiment

1) When the parameters ξ (damping ratio) and ω_n (damping natural frequency) are adjusted, the dynamic performance of the second order system will be difference.

2) Let students master the test method about the second order system dynamic performance

(2) The equipment of plant and controller

Hard ware: POFM - Arm, and PC.

Software: C ++ , or MATLAB

(3) Content of Experiment

Calculate the parameters of second order motor ω_n, ξ. Give out the step input, and adjust the damping ratio ξ, observe and record the response curve, fill the curve into the table 2.2. And then, test and record the percent overshoot $P.O.\%$, peak value time t_p, Settling time t_s, steady state error $e_{ss}(t)$, fill them into table 2.3.

2.2.2 电动机的阶跃响应实验

(1) 实验目的

1) 通过调节参数 ξ (阻尼比), ω_n (无阻尼自然频率), 可以看出二阶系统的动态性能变化。

2) 让学生掌握二阶系统动态性能的测试方法。

(2) 实验设备

硬件设备：单自由度便携式机械臂，计算机。

软件环境：C ++ ，或者 MATLAB；

(3) 实验内容

计算电机模型二阶系统的 ω_n，ξ。首先给定单位阶跃信号，改变二阶系统的阻尼比 ξ，观察并记录输出的响应曲线填写到表 2.2，然后测量并记录超调量 $P.O.\%$，峰值时间 t_p 和调节时间 t_s，稳态误差 $e_{ss}(t)$ 等性能指标值，填写到表 2.3。

Table 2.2 The Experiment Response Curve of Adjust Damping Ratio ξ

表 2.2 阻尼比调整的实验响应曲线

阻尼比 ξ	$0<\xi<1$	$\xi=1$	$\xi>1$
阶跃响应曲线			

Table 2.3 The Performance Index of Second Order System

表 2.3 二阶系统的性能指标

阻尼比 ξ	超调量 $P.O.\%$	峰值时间 t_p	调节时间 t_s	稳态误差 $e_{ss}(t)$
$0<\xi<1$				
$\xi=1$				
$\xi>1$				

2.3 Experiment Instruction

2.3 实验指导书

姓名（Name）:						院系（School）:			
专业（Major）:						学号（Student ID）:			
起至日期：	年	月	日	至	年	月	日		
from:	Y	M	D	to	Y	M	D		

实验题目：二阶系统建模实验

Experiment Title: Modeling of Second - order System

Ⅰ. Experimental Purpose

1. Through experiments to understand the affection of the dynamic properties in second - order system when parameters ζ (damping ratio) and ω_n (damped natural frequency) Vary;

2. Study the dynamic performance test method for second - order system.

Ⅱ. Experimental Device

Hardware device: A portable one - freedom robot arm, a computer (upper monitor);

Software: VC + + SDE (Software Development Environment).

Ⅲ. Experimental Content

1. Calculate the parameters ω_n, ζ in model of second - order system, change the damping ratio of the second - order system according to given unit step signal, observe and record the output of the response curve;

2. Identify the parameters of system and give out the transfer function of system G(s);

3. According to the G(s), write the unit step response program with MATLAB and do experiment in simulation;

4. Keep step signal unchanged, change the damping ratio of second - order system. observe and record overshoot *P. O.* %, peak time t_p, setting time t_s, steady state error $e_{ss}(t)$;

5. Analyze the results of experiment.

一、实验目的

1. 通过实验了解参数 ζ（阻尼比），ω_n（阻尼自然频率）的变化对二阶系统动态性能的影响；

2. 掌握二阶系统动态性能的测试方法。

二、实验设备

硬件设备：单自由度便携式机械臂一台，计算机（上位机）一台；

软件设备：VC ++ 开发环境。

三、实验内容

1. 计算电动机模型二阶系统的 ω_n，ξ，给定单位阶跃信号，改变二阶系统的阻尼比，观察并记录输出的响应曲线；

2. 运用最小二乘法进行系统参数辨识，求取传递函数；

3. 根据被控对象的传递函数，用 MATLAB 编写单位阶跃响应程序并进行实验；

4. 阶跃信号不变，改变二阶系统的阻尼比，测量并记录超调量 *P. O.* %，峰值时间 t_p 和调节时间 t_s，稳态误差 $e_{ss}(t)$；

5. 实验结果分析。

2.4 Requirements for The Experiment Report

1) Record the step response curves and performance index with different damping ratio ζ, and analyze the affection for system with changes of parameter ζ, sum up the regulation;

2) Compare the experimental results with the theoretical value, and analyze the characteristic and performance in DC motor step response.

2.4 实验报告要求

1）记录不同阻尼比 ζ 的阶跃响应曲线和性能指标值，分析参数 ζ 变化对系统的影响，总结规律；

2）将实验结果与理论值进行比较分析，分析电机二阶系统阶跃响应特点和性能。

2.5 Experiment Notes

1) To preview before the experiment.

2) Be care of laboratory equipment.

2.5 实验注意事项

1）做实验之前要预习

2）爱护实验设备

Appendix:

Program of LMS parameter identification in MATLAB

According to table 2.1, use Least Mean Square algorithm to identify the parameters of a and b, the program in MATLAB is shown as following:

附录:

运用 MATLAB 实现最小二乘参数辨识程序

按表 2.1 实验数据，利用最小二乘法原理确定 a，b 这两个参数，MATLAB 实现程序如下：

```
%表1  原始数据(original data)
close all;
    t=[0.036 0.196 0.286 0.375 0.464 0.571 0.714 0.768];
    % y=[0 1.047 2.094 3.142 4.189 5.236 6.283 6.667];
    y=[0 60 120 180 240 300 360 382];
    % y=y/(6.67);
    y=log(383 - y);
%线性拟合 (linear approximation)
    n=7;Lx=0;Ly=0;Lxx=0;Lxy=0;Lyy=0;
```

```
    for i = 1 : n
    Lx = Lx + t(i);
    Ly = Ly + y(i);
    Lxx = Lxx + t(i) * t(i);
    Lxy = Lxy + t(i) * y(i);
    Lyy = Lyy + y(i) * y(i);
    end
    a = (n * Lxy -  Ly *  Lx)/(n *  Lxx - Lx * Lx);
    b = (Ly - a * Lx)/n;
% 参数和模型求取(Find out parameters and Model)
    w1 = -2.303 * a;
    w2 = w1/(1 - 10^(-b));
    T = 1/sqrt(w1 *  w2)
    E = w1 *  w2 * T
    z = [];
    p = [ -w1 -w2];
    k = w1 * w2;
    sys = zpk(z,p,k)
% 图形绘制 (Drawing curve)
    figure(1)
    plot(t,a * t + b,'r',t,y)% 拟合前后直线对比 (compare with original curve)
    figure(2)
    step (sys)% 阶跃响应曲线 (step response curve )
```

计算结果如下 (results of calculation):

T=0.1088, ξ=9.1880, 进一步可以求得系统特征根 (then the roots of system will be obtained):

$$\begin{cases} s_1 - \dfrac{1}{T}[\xi - \sqrt{\xi^2 - 1}] \\ s_2 = \dfrac{1}{T}[\xi + \sqrt{\xi^2 - 1}] \end{cases}$$ 。

Chapter 3　Root Locus Compensation Experiments for Control System

第3章　控制系统的根轨迹校正实验

For the POFR – Arm system, by means of the root locus method, we can analyze its stability and dynamic response performance. And a compensator should be designed by root locus method to improve the system performance also. The block diagram of compensation control system is shown in Figure 3. 1.

对于便携式单自由度机械臂这一被控对象，可以用根轨迹的方法分析其稳定性和动态响应性能。同时，也可以运用根轨迹的方法设计一个校正装置，来改善系统的性能，其控制系统框图如图3. 1所示。

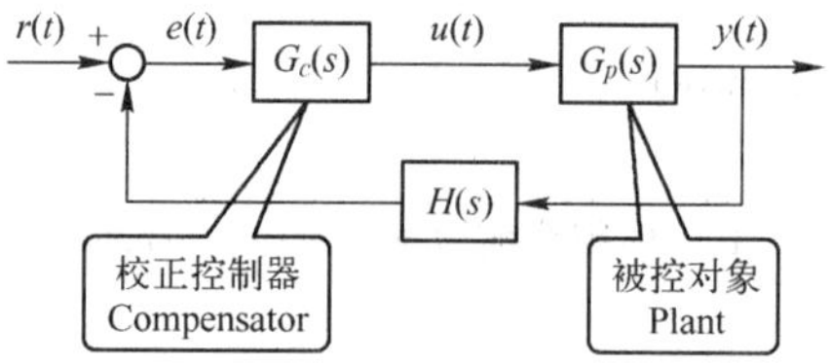

Figure. 3. 1　Feedback Control System Diagram with Compensator

图3. 1　反馈控制系统框图

In this chapter, we will introduce how to design a compensator by root locus method. Students will do the compensator design experiment and verify that the system performance is better than without compensator.

本章将介绍如何运用根轨迹的方法进行系统校正，并且用实验验证校正后的系统的性能比校正前的系统性能是否有所改善。

3. 1　Analysis for the Transfer Function and Root Locus of The Plant

3. 1　被控对象的传递函数与根轨迹图分析

Root locus Means the closed – loop characteristic square's root locus in s plane when gain K change from zero to positive infinity.

根轨迹是指系统的增益 K 由零到正无穷大时的闭环特征方程的根在 s 平面上的变化轨迹。在系统的增益 K 由零变

When gain K from zero to negative infinity called complement root locus. And it is called full root locus when K changes from negative infinity to positive infinity. Normally, root locus is considered as the gain K varies from zero to positive infinity.

According to the conclusion in Chapter 2, the transfer function of DC motor is as following:

$$G(s)=\frac{84.42}{s(s+18.376)} \tag{3.1}$$

Assume that gain K in (3.1) is adjustable, and the formula can be written as (3.2):

$$G(s)=\frac{K}{s(s+18.376)} \tag{3.2}$$

The structure diagram of compensation feedback control system is shown in Figure 3.2. Using MATLAB tools, the root locus diagram in Figure 3.3 should be drawn when K selected different data (K is changed from zero to positive infinity).

化到负无穷大时闭环特征方程的根的轨迹为补根轨迹。系统闭环特征方程的根轨迹与补根轨迹称为全根轨迹。通常情况下根轨迹是指增益 K 由零到正无穷大下的根的轨迹。

根据第 2 章的分析，被控对象的传递函数为：

$$G(s)=\frac{84.42}{s(s+18.376)} \tag{3.1}$$

假定式（3.1）的增益 K 是可变化（可调节的）的，则式（3.1）可以写为式（3.2）：

$$G(s)=\frac{K}{s(s+18.376)} \tag{3.2}$$

校正反馈控制系统的实物构成如图 3.2 所示。运用 MATLAB 工具可以很容易地画出式（3.2）中当 K 从零变化到无穷大时的根轨迹图，如图 3.3 所示：

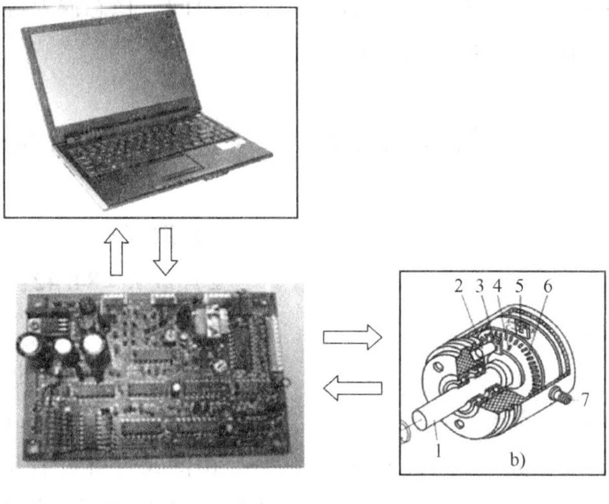

Figure 3.2 The Structure Diagram of Compensation Feedback Control System

图 3.2 校正反馈控制系统的实物构成示意图

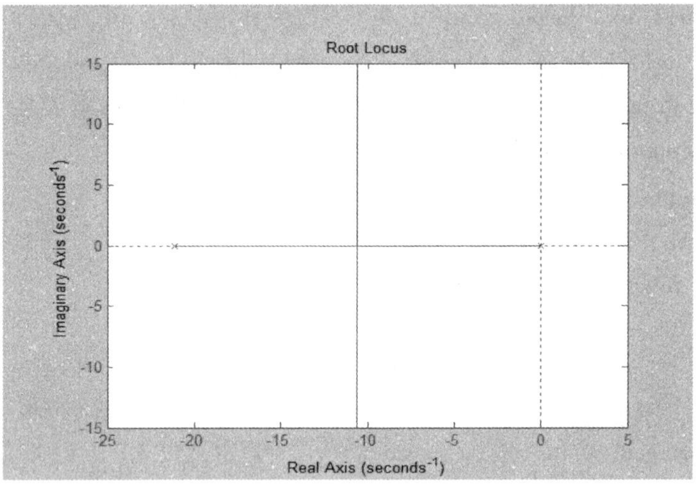

Figure 3.3 The Root Locus of Plant (POFR – Arm)

图 3.3 被控对象（便携式单自由度机械臂）的根轨迹图

It is obviously that the POFR – Arm is a stabilizing system, that is to say, no matter how K changes, the root of the closed – loop system should not fall in right s plane and cause instability. By adjusting the parameter of K, we can observe how it affects system performance, that is, adjusting the parameter of K will improve the values of system overshoot, setting time, steady state error etc.

由这个系统可以看出，被控对象单自由度机械臂是一个自身稳定的系统，也就是说，无论 K 怎么变化，都不会导致闭环系统的根落入右半 s 平面而引起不稳定。我们可以通过调整 K 的变化观察其对系统性能指标的影响。即，调节 K 值的大小，来改善系统的超调量、过渡过程时间、稳态误差等值。

The following three experiments should be done:

1) Take $K=1$;

2) Take $K=84.42$;

3) Take $K=200$。

进行以下 3 个实验：

1) 取 $K=1$;

2) 取 $K=84.42$;

3) 取 $K=200$。

Draw the step response curve as Figure 3.4, Figure 3.5, Figure 3.6.

分别画出对应的阶跃响应曲线如图 3.4、图 3.5、图 3.6 所示。

It can be seen in Figure 3.4, when $K=1$, the POFR – Arm is a system without overshoot, steady – state error tends to zero.

由图 3.4 可以看出，当 $K=1$，单自由度机械臂的系统没有超调，稳态误差趋向于零。

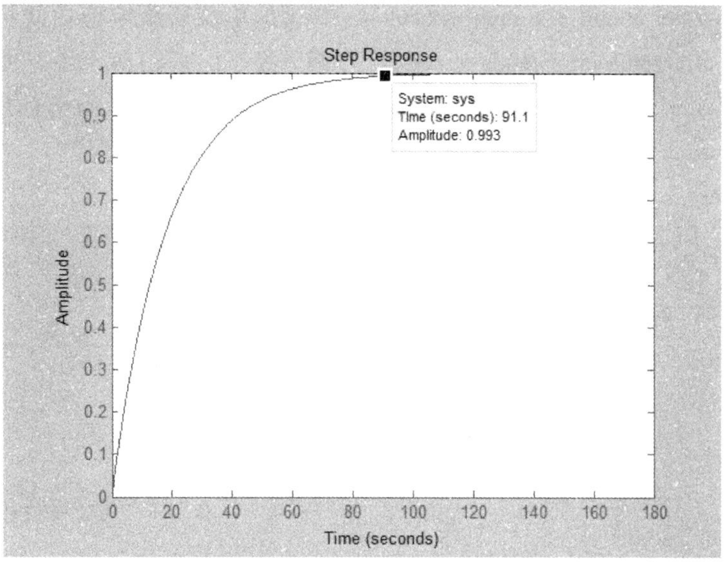

Figure 3.4 Step Response Curve of POFR - Arm when $K=1$

图 3.4 $K=1$ 时单自由度机械臂的单位阶跃响应曲线

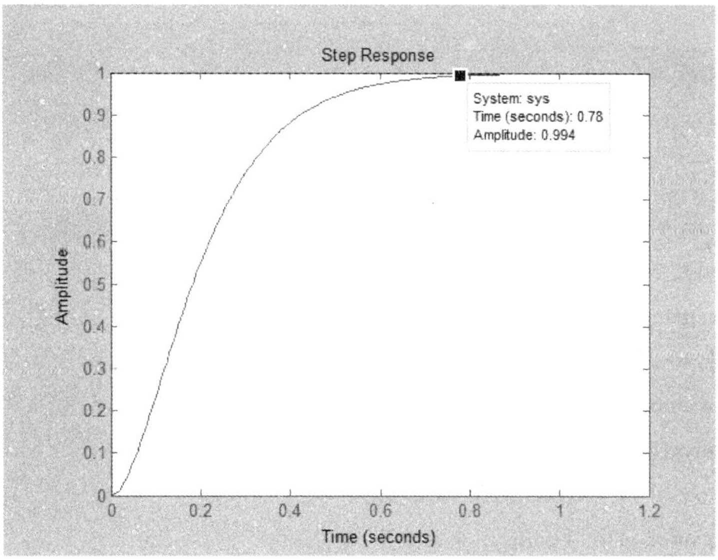

Figure 3.5 Step Response Curve of POFR - Arm when $K=84.42$

图 3.5 $K=84.42$ 时单自由度机械臂的单位阶跃响应曲线

Let $K = 84.42$, the system response curve can be obtained as shown in Figure 3.5.

取 $K = 84.42$，系统的响应曲线如图 3.5 所示。

It can be seen when $K = 84.42$, the POFR - Arm system is still a system without overshoot; steady - state error tends to zero, but response faster than Figure 3.4 (Scale interval in Figure 3.5 is 0.2 seconds, Scale interval in Figure 3.4 is 20 seconds).

当 $K = 84.42$，可以看出一自由度机械臂的单位阶跃响应曲线依然是没有超调，稳态误差趋向于零。但响应速度比图 3.4 快（图 3.5 的坐标尺度为 0.2 s，而 3.4 的坐标尺度为 20 s）。

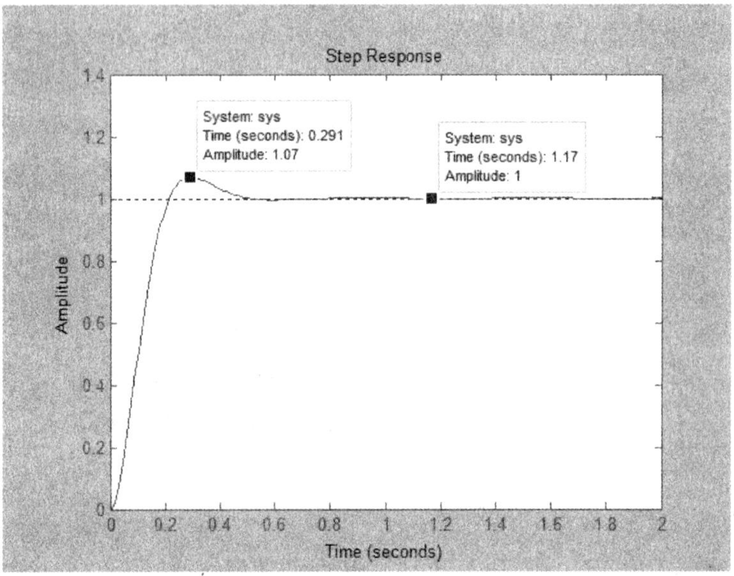

Figure 3.6 Step Response Curve of POFR - Arm when $K = 200$

图 3.6 $K = 200$ 单自由度机械臂的单位阶跃响应曲线

It can be seen in Figure 3.6, when $K = 200$, the POFR - Arm system is a system with 0.0731 percent overshoot, steady - state error tends to zero, but response faster than Figure 3.4 and Figure 3.5 (The horizontal scale interval in Figure 3.6 is 0.2 seconds, scale interval in Figure 3.5 is 0.2 seconds, scale interval in Figure 3.4 is 20 seconds).

从图 3.6 可以看出，当 $K = 200$，一自由度机械臂系统的超调量近似为 0.0731，稳态误差最终趋向于零，但响应速度比图 3.4 和图 3.5 更快（注意横坐标的尺度，图 3.6 为 0.2 s；图 3.5 为 0.2 s，图 3.4 为 20 s）。

It is obviously in the curves, when K takes different values, it will cause the different performance in control system. There is MATLAB programming shown in Appendix 3.1.

由上面的图形可以看出，当 K 取不同数值时，将会导致不同的控制性能指标。附录 3.1 是关于 MATLAB 的程序实现。

3.2 System Design and Experiment of Root Locus Lead Compensation

Lead compensator based on root locus is designed by adding a series component, namely by adding the open - loop poles and zeros to adjust the original system, which will obtain the desired root locus. The transfer function of compensation is usually taken as following formula:

$$Gc(s) = k\frac{s + Z_D}{s + P_D} \quad (3.3)$$

Where $|Z_D < P_D|$;

The main idea of compensation by root locus is to assume that the closed - loop control system has a pair of dominant poles, thus the dynamic performance of the system should be approximately equal to second - order system described by the dominant poles.

(1) Calculate damping coefficient ξ and natural oscillation frequency ω_n according to expected performance index.

(2) Calculate the pair of dominant poles:

$$s_{1,2} = \xi\omega_n \pm \omega_n \sqrt{1 - \xi^2}$$

here, the approximate dominant pole is

$$s_{1,2} = -6 \pm 5.3j$$

(3) Determine zeros and poles of the lead compensator (It shown in Figure 3.7):

1) Add directly a real zero of phase advance network underneath the expected close - loop poles. Take $z = -6$.

3.2 根轨迹超前校正系统设计与实验

基于根轨迹的超前校正设计是通过增加一个串联超前校正环节，即增加开环极点和零点，对原来的系统进行调整，得到所需的根轨迹。校正环节的传递函数通常取下面的公式实现：

$$Gc(s) = k\frac{s + Z_D}{s + P_D} \quad (3.3)$$

其中$|Z_D| < |P_D|$；

用根轨迹校正的基本思想就是假设控制系统有一对闭环主导极点，这样系统的动态性能就可以近似地用这对主导极点所描述的二阶系统来表征。

(1) 根据期望的性能指标计算出阻尼系数ξ和自然振荡频率ω_n。

(2) 计算一对主导极点：

$$s_{1,2} = \xi\omega_n \pm \omega_n \sqrt{1 - \xi^2}$$

这里求得近似主导极点为

$$s_{1,2} = -6 \pm 5.3j$$

(3) 确定超前校正器的零点和极点（如图 3.7 所示）：

1) 直接在期望的闭环极点位置下方增加一个相位超前网络的实零点。这里取$z = -6$。

2) Using phase angle of pole in the compensator network, make sure that the expected dominant pole satisfy phase angle conditions in the root locus:

$$\angle G(s) = \sum_{i=1}^{m} \angle(s - z_i) - \sum_{j=1}^{n} \angle(s - p_j)$$
$$= \sum_{i=1}^{m} \varphi_i - \sum_{j=1}^{n} \theta_j = (2k+1)\pi$$

Where $\sum \theta_j$ is algebraic sum of phase angle of vector from all other open - loop poles to pole in question, $\sum \varphi_i$ is algebraic sum of phase angles of vectors from all other open - loop zeros to the pole.

3) According to the amplitude condition , we could obtain K:

$$K \frac{\prod_{i=1}^{m} |(s - z_i)|}{\prod_{j=1}^{n} |(s - p_j)|} = 1$$

2）利用校正网络极点的相角，使得系统在期望主导极点上满足根轨迹的相角条件：

$$\angle G(s) = \sum_{i=1}^{m} \angle(s - z_i) - \sum_{j=1}^{n} \angle(s - p_j)$$
$$= \sum_{i=1}^{m} \varphi_i - \sum_{j=1}^{n} \theta_j = (2k+1)\pi$$

式中，$\sum \theta_j$ 为所有各其他开环极点到该极点的向量的相位夹角之代数和，$\sum \varphi_i$ 为所有开环零点到该极点的向量的相位夹角之代数和。

3）根据幅值条件得到 K：

$$K \frac{\prod_{i=1}^{m} |(s - z_i)|}{\prod_{j=1}^{n} |(s - p_j)|} = 1$$

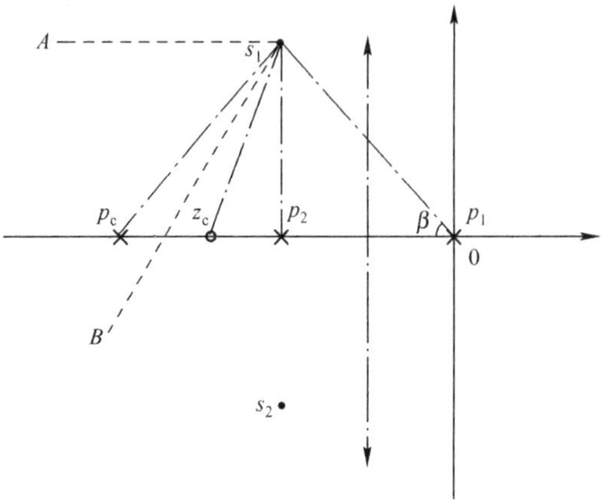

Figure 3.7 Determine Zeros and Poles of the Lead Compensator

图 3.7 超前校正零极点的确定

According to the MATLAB programming (See attached Appendix 3.2), we get some parameters as following:

$t_p = 0.0730$;

由 MATLAB 编程实现（见附录 3.2），我们可以得到下面的参数：

$t_p = 0.0730$;

$y_{ss} = 1.0000$;

$ct = 0.064$;

$t_s = 0.3892$;

$s_p = 102.2$;

$k = 26.0853$.

Transfer function of the compensator is:

$$G_c = 26.0853 * \frac{s+6}{s+102.2} \quad (3.4)$$

That is: pole $s_p = 102.2$, parameter $K = 26.0853$, overshoot $P.O.\% = 6.84\%$, Setting time $t_s = 0.3892$. all meet the performance requirement.

Root locus after compensation is shown in Figure 3.8, System step response after compensation is shown in Figure 3.9:

$y_{ss} = 1.0000$;

$ct = 0.064$;

$t_s = 0.3892$;

$s_p = 102.2$;

$k = 26.0853$。

综上得到校正网络传递函数为:

$$G_c = 26.0853 * \frac{s+6}{s+102.2} \quad (3.4)$$

即极点为 $s_p = 102.2$，系数 k = 26.0853，超调量 $P.O.\% = 6.84\%$，调节时间为 $t_s = 0.3892$，满足性能指标要求。

校正后的根轨迹如图 3.8 所示，阶跃响应曲线如图 3.9 所示。

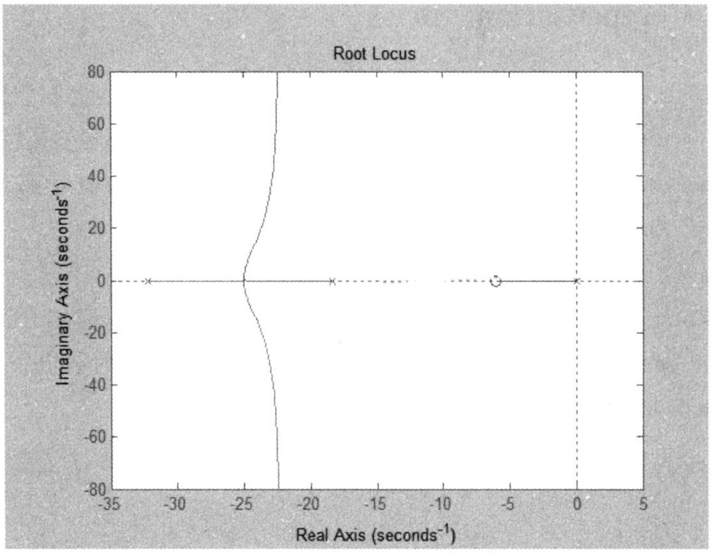

Figure 3.8 Root Locus after Compensation

图 3.8 校正后的根轨迹图

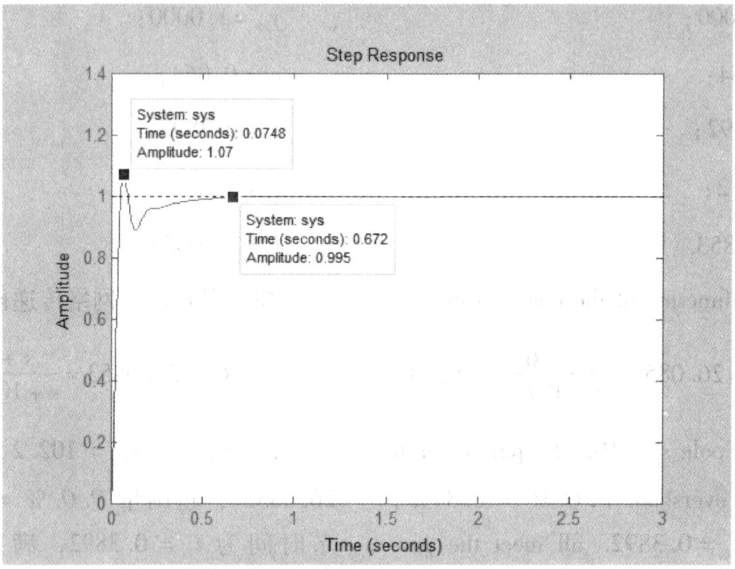

Figure 3.9 System Unit Step Response after Compensation

图 3.9 校正后的单位阶跃响应曲线

3.3 System Design and Experiment of Root Locus Lag Compensation

3.3 根轨迹滞后校正设计与实验

If the system cannot fit the steady - state performance requirements, we can design a lag compensator to increase the steady - state accuracy, while keep its required dynamic performance.

若系统稳态性能不能满足要求，也可以采用串联滞后校正以提高系统的稳态精度，同时保持其动态性能满足性能指标要求。

Lag compensator increases the system gain by adding open - loop dipole. The transfer function of lag compensator is:

滞后校正采用增加开环偶极子来增大系统增益。滞后校正网络的传递函数为

$$G_c(s) = k_0 \frac{s + z_c}{s + p_c} \tag{3.5}$$

Where $|z_c| > |p_c|$.

其中$|z_c| > |p_c|$。

After knowing open loop transfer function

已知系统的开环传递函数为

$$G(s) = \frac{84.42}{s(s + 18.376)}$$

the required overshoot $P.O.\% < 10\%$, setting time $t_s < 0.5$ s, the steady - state error of unit ramp response $e_{ss} \leqslant 0.1$.

The open loop transfer function after lag compensation is:

$$G_cG(s) = \frac{84.42}{s(s+18.376)} \times k_0 \times \frac{s+z}{s+p} = \frac{84.42/18.376}{s(1/18.376s+1)} \times k_0 \times \frac{s+z}{s+p}, \quad (3.6)$$

As the steady - state error of unit ramp response is:

$$e_{ss} = \frac{1}{(84.42/18.376)k_0} \leqslant 0.1,$$

so $k_0 \geqslant 2.176$. As requested, the ratio between zero and pole of dipole should be 10, so we could choose $z = 0.15$, $p = 0.015$. In determining the pole and zero value, the experimental method can be used to obtain the compensator transfer function as shown in formula (3.7). And the programming of experimental of determine parameters in lag compensator is shown in Appendix 3.3.

$$G_c = 2.5 \times \frac{s+0.15}{s+0.015} \quad (3.7)$$

That is: overshoot $P.O.\% = 8.16\%$, setting time $t_s = 0.4141$ s, steady - state error $e_{ss} = 0.0077 < 0.1$. All meet the required performance index.

The root locus after lag compensation for POFR - Arm is shown in figure 3.10, and its unit step response curve is shown in figure 3.11.

Compensators which above mentioned should be done both by simulation and by RC circuit.

如果要求其超调量 $P.O.\% < 10\%$，调节时间 $t_s < 0.5$ 秒，单位斜坡响应的稳态误差是 $e_{ss} \leqslant 0.1$。

校正后的系统开环传递函数为：

$$G_cG(s) = \frac{84.42}{s(s+18.376)} \times k_0 \times \frac{s+z}{s+p} = \frac{84.42/18.376}{s(1/18.376s+1)} \times k_0 \times \frac{s+z}{s+p}, \quad (3.6)$$

由单位斜坡响应稳态误差为：

$$e_{ss} = \frac{1}{(84.42/18.376)k_0} \leqslant 0.1,$$

可得 $k_0 \geqslant 2.176$。按照要求，偶极子的零点和极点比值应为 10，因此我们可以选择 $z = 0.15$，$p = 0.015$。在确定零极点值时，可采用实验的方法，获得校正环节的传递函数如式（3.7）所示。滞后实验方法确定参数的程序如附录 3.3 所示。

$$G_c = 2.5 \times \frac{s+0.15}{s+0.015} \quad (3.7)$$

即超调量为 $P.O.\% = 8.16\%$，调节时间 $t_s = 0.4141$ s，稳态误差 $e_{ss} = 0.0077 < 0.1$，满足性能指标要求。

校正后的单自由度机械臂系统的根轨迹如图 3.10 所示：校正后的系统阶跃响应如图 3.11 所示。

本文以上介绍的校正环节可通过软件仿真实现，也可以通过 RC 电路实现。

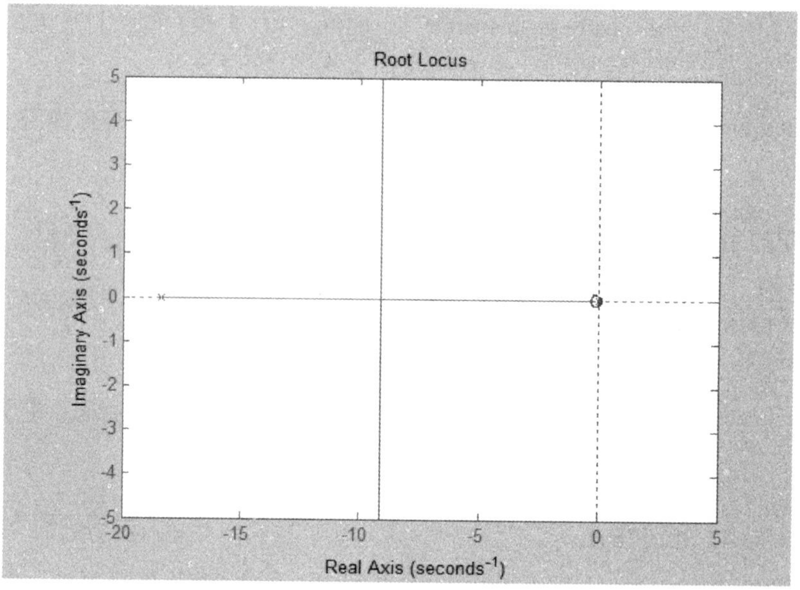

Figure 3. 10 The Root Locus after Lag Compensation

图 3. 10 滞后校正后根轨迹

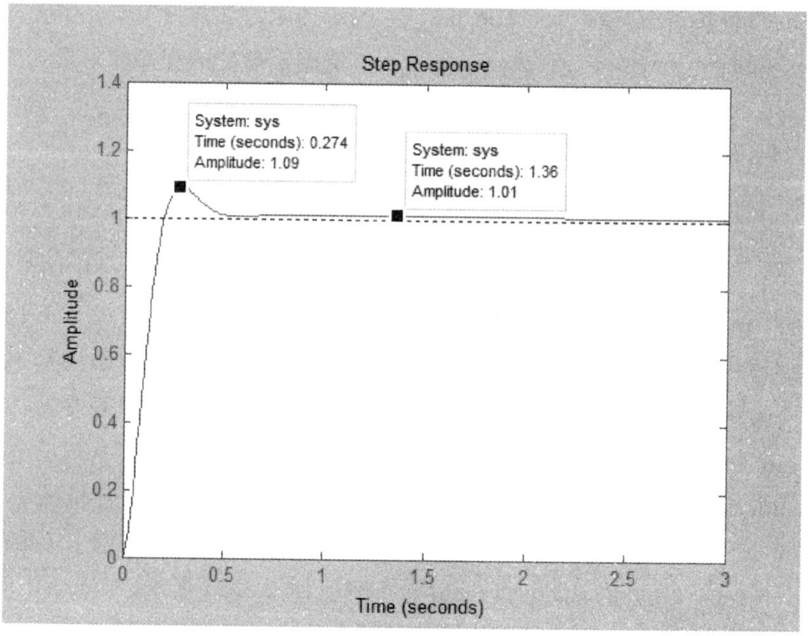

Figure 3. 11 Unit Step Response Curve after Lag Compensation

图 3. 11 滞后校正后系统阶跃响应曲线

3.4 Experiment Instruction

3.4 实验指导书

姓名（Name）： 院系（School）：

专业（Major）： 学号（Student ID）：

起至日期： 年 月 日 至 年 月 日

from： Y M D to Y M D

实验题目：根轨迹法校正实验

Experimental Title：Root Locus Compensation Experiment

Ⅰ. Known Transfer Function and Parameter

POFR－Arm is the controlled object，transfer function：

$$G(s)=\frac{4.587}{s(0.054s+1)}$$

Ⅱ. Experimental Requirements：

1. Finish the close－loop response curve when $K=1$，100，200；
2. Finish the lead compensation experiment；
3. Finish the lag compensation experiment；
4. Compare the difference between advance and lag compensation experiment.

Ⅲ. Experiment Content

1. According to the transfer function of controlled object，MATLAB write corresponding program in MATLAB and test it；
2. According to the requirement，design transfer function of advance and lag compensation；
3. Connect computer and controlled object，do the experiment，and save the root locus and step response curve after compensation.

Ⅳ. Print the Output Result

Ⅴ. Analyze the Result

一、已知技术参数与传递函数

实验单自由度机械臂为被控对象，传递函数为：

$$G(s)=\frac{4.587}{s(0.054s+1)}$$

二、实验要求

1. 完成 $K=1$，100，200 时系统的闭环响应曲线；
2. 完成超前校正实验；
3. 完成滞后校正实验；
4. 比较超前与滞后校正实验的区别。

三、实验内容

1. 根据被控对象的传递函数，运用 MATLAB 编写根轨迹绘制程序并进行实验；
2. 按照要求设计超前、滞后校正传递函数；
3. 连接计算机与被控对象，进行根轨迹校正实验，并保存校正后系统的根轨迹图与单位阶跃响应图。

四、打印实验输出结果

五、实验结果分析

3.5 Requirements of The Experiment Report

3.5 实验报告要求

1） Complete the close－loop response curve when $K=1$，100，200.

1）完成 $K=1$，100，200 时系统的闭环响应曲线；

2) Complete the lead compensation experiment.

2) 完成超前校正实验;

3) Complete the lag compensation experiment.

3) 完成滞后校正实验;

4) Compare the difference between lead and lag compensation experiment.

4) 比较超前与滞后校正实验的区别。

附录 3.1 测试计算被控对象的闭环响应
Appendix 3.1 Analyze the closed loop response of controlled abject

```
num1 = [84.42];
den1 = [1 18.376 0];
G = tf(num1,den1);
sys = feedback(G,1);
figure(1);
[y,t,x] = step(sys,3);
step(sys);
mp = max(y);
tp = spline(y,t,mp)
cs = length(t);
yss = y(cs)
ct = (mp - yss)/yss
```

附录 3.2 根轨迹超前校正
Appendix 3.2 System design of root locus lead compensation

```
den1 = [1 18.376 0];
G = tf(84.42,den1);
x = -1:-0.01:-20;
s = -6 +5.3 * j;
Zlead = 6;
angs = 90 - 2 * angle(s - 0) * 180/pi - angle(s - x) * 180/pi;
Plead = spline(angs,x, -180)
k = abs((s * (s + 18.376) * (s + Plead))/(84.42 * (s + Zlead)))
  D = zpk( -Zlead,Plead,1)
  G0 = tf(k * 84.42 * [1 Zlead],[1 50.61 592.26  0]);
  figure(1);
  rlocus(G0)
  sys = feedback(G0,1);
  figure(2);
```

```
step(sys,3)
[y,t,x] = step(sys,3);
  mp = max(y);
  tp = spline(y,t,mp)
  cs = length(t);
  yss = y(cs)
  ct = (mp - yss)/yss
  L = length(t);
while(y(L) >0.98 * yss)&(y(L) <1.02 * yss)
L = L - 1;end
  ts = t(L) %求调节时间
```

附录 3.3 根轨迹滞后校正

Appendix 3.3 System design of root locus lag compensation

```
num1 = [84.42];
  den1 = [1 18.376 0];
  G = tf(num1,den1);
  num2 = [1 0.15];
  den2 = [1 0.015];
  Gc = tf(num2,den2);
% num = [1 7];
% den = [1 60.5 18.376 1111];
  H = tf(2.5 * G * Gc);
  sys = feedback(H,1);
  figure(1);
  rlocus(H)
  figure(2);
  step(sys,3)
  [y,t,x] = step(sys,3);
    mp = max(y);
    tp = spline(y,t,mp)
    cs = length(t);
    yss = y(cs)
    ct = (mp - yss)/yss
    L = length(t);
  while(y(L) >0.98 * yss)&(y(L) <1.02 * yss)
  L = L - 1;end
  ts = t(L) %求调节时间
```

Chapter 4　Frequency Compensation Experiment

第4章　频率法校正实验

Use frequency method to design a compensator, mainly by changing the shape of frequency characteristics curves, to make sure that it has suitable high - frequency, intermediate frequency, low - frequency characteristics and stable margin, so that the system will give out closed - loop output of high quality. The main frequency index of performance are adopted as phase margin, gain margin, the resonance peak and the frequency bandwidth, instead of percent overshoot, settling time etc. , which are used in time domain. Therefore, the frequency method is an indirect method for control system design and analysis. A useful tool is Bode diagram analysis method.

用频率法校正设计校正装置，主要是改变频率特性曲线的形状，使之具有合适的高频、中频、低频特性和稳定裕度，以得到满意的闭环品质。用频率法校正控制系统时，通常是以频域指标如相位裕度、增益裕度、谐振峰值和频带宽度来衡量和调整控制系统动态响应性能，而不是按照时域指标如超调量、调节时间和稳态误差等来进行。所以，频率法是一种间接的控制系统设计与分析方法，该方法所利用的工具为伯德图分析法。

In this chapter, POFR - Arm is used as the controlled plant, students should study how to design an compensator by the frequency method. Its block diagram is the same as in Figure 3. 1, and the real control system is shown as Figure 3. 2. Experiments in this chapter include two types of series compensator—— lead and lag compensation.

本章使用单自由度机械臂为被控对象，学生们可学习如何运用频率法设计校正装置。其控制系统方框图的构成与图3. 1是相同的，实物图的构成则与图3. 2相同。本章实验包括两类串联校正系统——超前和滞后校正。

4. 1　Experiment of Series Lead Compensation

4. 1　串联超前校正实验

According to the analysis in chapter 2, the open - loop transfer function of POFR - Arm is:

根据第2章的分析，单自由度机械臂系统的开环传递函数为

$$G(s)=\frac{4.587}{s(0.054s+1)} \quad (4.1)$$

If the required index of velocity error constant is $K_v \geqslant 10s^{-1}$, the phase margin $\gamma(\omega_c)\geqslant 50°$, the lead compensator can be used to design a new control system.

Write (4.2) in the frequency form:

$$G(j\omega)=\frac{4.587}{j\omega(0.054j\omega+1)} \quad (4.2)$$

The step of lead compensator experiment designing is as following:

Write the compensator transfer function:

$$G_c=\frac{s/\omega_1+1}{s/\omega_2+1}=\frac{s/T+1}{s/\frac{\gamma_d}{T}} \quad (4.3)$$

where ω_1, ω_2 are crossover frequency.

1) Determine the open loop gain according to the requirements for the steady state error, $K=100$, and the $k_c=100/4.587=21.8$.

2) According to the given phase margin, estimate the additional phase angle position.

$$A(\omega_c)\approx\frac{100}{\omega_c\cdot 0.054\omega_c}=1 \quad (4.4)$$

So we obtain $\omega_c=43$, the phase margin is

$$\gamma=180°+(-90-\arctan 0.054\omega_c)$$
$$\approx 23.299°,$$

it is obviously that the γ cannot meet the requirements.

Firstly, the system Bode diagram is drawn in Figure 4.1.

As is shown in Figure 4.1, the phase margin is 24.3°, the steady state velocity error constant is 100.

$$G(s)=\frac{4.587}{s(0.054s+1)} \quad (4.1)$$

要求校正后的系统稳态速度误差系数 $K_v \geqslant 10s^{-1}$，相位裕度 $\gamma(\omega_c)\geqslant 50°$，我们便可以用超前校正的方法，来设计一个新的控制系统。

将式（4.1）写成频率特性的形式：

$$G(j\omega)=\frac{4.587}{j\omega(0.054j\omega+1)} \quad (4.2)$$

采用超前校正的实验步骤如下：

写出校正环节的传递函数为：

$$G_c=\frac{s/\omega_1+1}{s/\omega_2+1}=\frac{s/T+1}{s/\frac{\gamma_d}{T}} \quad (4.3)$$

其中 ω_1，ω_2 为交接频率。

1）根据稳态误差的要求确定系统的开环放大系数，$K=100$，$k_c=100/4.587=21.8$。

2）根据给定相位裕度，估计出需要附加的相角位移。

$$A(\omega_c)\approx\frac{100}{\omega_c\cdot 0.054\omega_c}=1 \quad (4.4)$$

故可得 $\omega_c=43$，相角裕度为

$$\gamma=180°+(-90-\arctan 0.054\omega_c)$$
$$\approx 23.299°,$$

可见相位裕度不满足要求。

首先画出系统的伯德图，如图 4.1 所示。

由图 4.1 可以看出，该机械臂的相角裕度是 24.3°，该系统的稳态速度误差系数为 100。

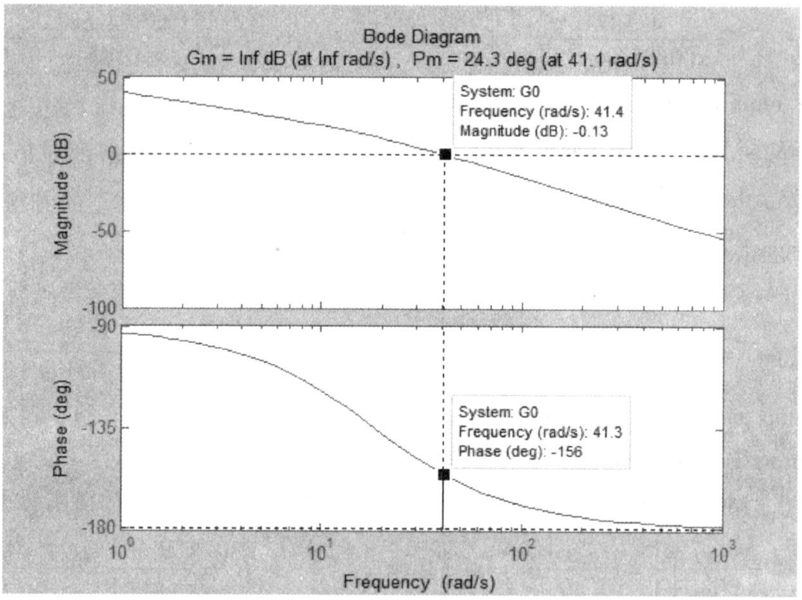

Figure 4.1　Bode Diagram of POFR – Arm

图 4.1　单自由度机械臂的伯德图

The unit step response of the system before compensation is shown in Figure 4.2:

系统校正前的单位阶跃响应如图 4.2 所示：

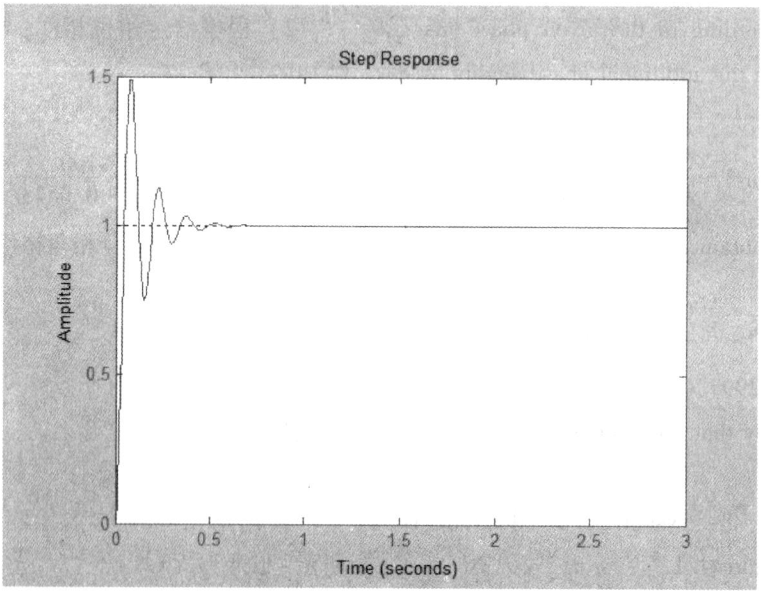

Figure 4.2　POFR – Arm's Step Response Curve before Compensation

图 4.2　校正前的单自由度机械臂的阶跃响应曲线

MATLAB code is shown in Appendix 4.1, and the constants will be obtained.

MATLAB 实现程序见附录 4.1，根据 MATLAB 程序可以计算得出下面的参数：

The peak margin before compensation is $h \to \infty$; The phase margin before compensation is $\gamma = 24.2619$ and the cross frequency is $\omega_c = 41.0867$.

未校正系统的幅值裕度为 $h \to \infty$；未校正系统的相角裕度为 $\gamma = 24.2619$；未校正系统的截止频率 $\omega_c = 41.0867$。

3) According to the system requirement for phase margin $\gamma(\omega_c) \geqslant 50°$, adding lead phase angle is: $\varphi_{\max} = 50° - 24° + 9° = 35°$, so $\varphi_{\max} = \arcsin \dfrac{\gamma_d - 1}{\gamma_d + 1} = 35°$.

Thus we will get $\gamma_d = 3.65$.

3）根据系统相位裕度要求 $\gamma(\omega_c) \geqslant 50°$，增加超前角为 $\varphi_{\max} = 50° - 24° + 9° = 35°$，则 $\varphi_{\max} = \arcsin \dfrac{\gamma_d - 1}{\gamma_d + 1} = 35°$。

可得 $\gamma_d = 3.65$。

4) Assume that the given cross frequency ω_c' is the geometric center of two crossover frequency ω_1 and ω_2, that is:

$$\omega_c' = \sqrt{\omega_1 \omega_2} = \omega_1 \sqrt{\gamma_d}$$

As $A(\omega_c) \approx 1$, so we get $\omega_1 = 34.14$, $\omega_2 = 113.69$, $\omega_c' = 59.5$.

4）如果系统设定校正后的穿越频率 ω_c' 为校正装置两交接频率 ω_1 和 ω_2 的几何中点，即

$$\omega_c' = \sqrt{\omega_1 \omega_2} = \omega_1 \sqrt{\gamma_d}$$

由 $A(\omega_c) \approx 1$，得 $\omega_1 = 34.14$，$\omega_2 = 113.69$，$\omega_c' = 59.5$。

5) The transfer function of compensator is:

$$G_c(s) = \frac{s/31.14 + 1}{s/113.69 + 1} \tag{4.5}$$

After compensation, the system open loop transfer function is:

$$G_c(s) = \frac{s/31.14 + 1}{s/113.69 + 1} \times \frac{100}{s(0.054s + 1)} \tag{4.6}$$

System Bode diagram after compensation is shown in figure 4.3.

The unit step response of the system after compensation is in Figure 4.4.

5）校正传递函数为：

$$G_c(s) = \frac{s/31.14 + 1}{s/113.69 + 1} \tag{4.5}$$

经过校正环节，系统的开环传递变为：

$$G_c(s) = \frac{s/31.14 + 1}{s/113.69 + 1} \times \frac{100}{s(0.054s + 1)} \tag{4.6}$$

系统校正后的伯德图如图 4.3 所示。

校正后的系统阶跃响应曲线如图 4.4 所示。

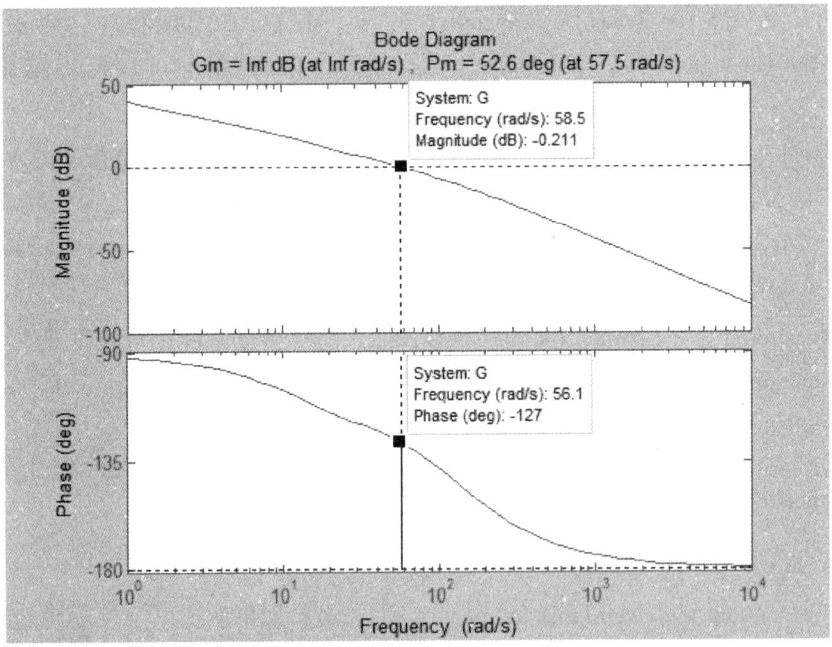

Figure 4. 3 POFM – Arm System Bode Diagram after Compensation

图 4. 3 校正后单自由度机械臂的伯德图

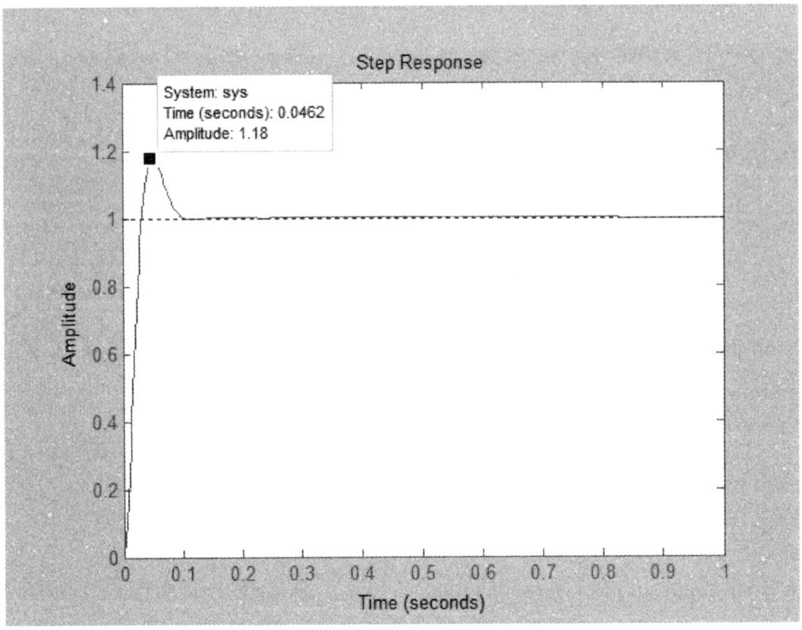

Figure 4. 4 POFM – Arm Step Response Curve after Compensation

图 4. 4 校正后的单自由度机械臂的阶跃响应曲线

Obviously:

The phase margin after compensation is $\gamma = 52.5963$; the cross frequency after compensation is $\omega_c = 57.4582$. So after compensation, the new system satisfies the requirements.

MATLAB programming code is shown in Appendix 4.2.

可见：

校正后系统的相角裕度为 $\gamma = 52.5963$；校正后系统的截止频率 $\omega_c = 57.4582$。即校正后的系统满足性能指标要求。

MATLAB 实现程序见附录 4.2。

4.2 Experiment of Series Lag Compensation

4.2 串联滞后校正实验

For the same plant of one freedom robot arm, requirements for the compensator are: system steady - state speed parameter $K_v \geqslant 100s^{-1}$, phase margin $\gamma(\omega_c) \geqslant 50°$, try determining the transfer function of lag compensator by experiment.

The transfer function of lag compensator:

$$W_c = \frac{s/\omega_2 + 1}{s/\omega_1 + 1} = \frac{Ts + 1}{\gamma_i Ts + 1} \quad (4.7)$$

where ω_1, ω_2 are crossover frequency.

General procedure of lag compensation is:

1) Determine the open loop gain according to the requirement for the steady state error, $K = 100$, $k_c = 100/4.587 = 21.8$.

2) According to the phase margin requirement $\gamma(\omega_c) \geqslant 50°$, and considering the phase lag in the vicinity of the crossover frequency caused by the impact of compensation device, increase appropriately compensation margin by 5° ~ 15°, we choose 10° here. So $\gamma(\omega_c) = 60°$, the through frequency after compensation is $\omega_c' = 10.8$ (See Figure 4.5).

同样针对单自由度机械臂系统，要求校正后的系统稳态速度参数 $K_v \geqslant 100s^{-1}$，相位裕度 $\gamma(\omega_c) \geqslant 50°$，试用实验方法确定滞后校正装置传递函数。

滞后校正装置传递函数为：

$$W_c = \frac{s/\omega_2 + 1}{s/\omega_1 + 1} = \frac{Ts + 1}{\gamma_i Ts + 1} \quad (4.7)$$

式中，ω_1，ω_2 为交接频率。

滞后校正的一般步骤如下：

1）根据稳态误差的要求确定系统的开环放大系数，$K = 100$，$k_c = 100/4.587 = 21.8$。

2）按照相位裕度要求 $\gamma(\omega_c) \geqslant 50°$，并考虑校正装置在穿越频率附近造成的相位滞后的影响，适当增加 5° ~ 15°的补偿裕度，这里增加 10°，故预选 $\gamma(\omega_c) = 60°$，对应的频率 $\omega_c' = 10.8$（如图 4.5 所示）为校正后的穿越频率。

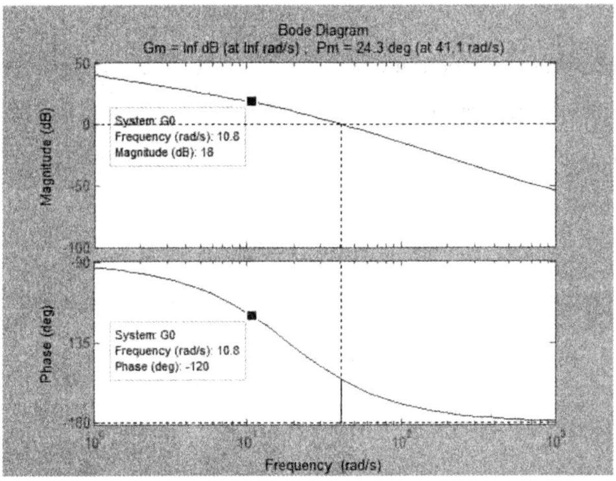

Figure 4. 5 System Bode Diagram of POFR – Arm before Compensation

图 4. 5 校正前单自由度机械臂的伯德图

3) Determine the former system frequency characteristic's decrement when the through frequency peak value drop to zero. Make the decrement equal to $-20\lg\gamma_i$, thus calculate γ_i. From the Bode diagram, we could obtain the Logarithm amplitude frequency characteristics gain: 18dB. So:

$20\lg\gamma_i = 18\text{ dB}$, $\gamma_i = 7.94$.

4) Choose the crossover frequency less than a decade frequency range, and then the other crossover frequency is obtained.

The pre – selection crossover frequency:

$$\omega_2 = \frac{\omega'_c}{8.0} = 1.35,\ \omega_1 = \frac{1}{\gamma_i T} = \frac{1.35}{7.94} = 0.17$$

So the transfer function of compensator is:

$$W_c = \frac{Ts+1}{\gamma_i Ts+1} = \frac{s/1.35+1}{s/0.17+1} \quad (4.8)$$

Bode Diagram of POFR – Arm after adding compensator is shown as Figure 4. 6.

3）确定原系统频率特性在穿越频率处的幅值下降到零分贝时所必需的衰减量。使这一衰减量等于 $-20\lg\gamma_i$，从而确定 γ_i 的值。从伯德图中查得对应穿越频率的对数幅频特性增益为 18dB，则得

$20\lg\gamma_i = 18\text{ dB}, \gamma_i = 7.94$。

4）选择一个交接频率低于穿越频率一倍到十倍频程，则另一交接频率也随之确定。

预选交接频率

$$\omega_2 = \frac{\omega'_c}{8.0} = 1.35, \omega_1 = \frac{1}{\gamma_i T} = \frac{1.35}{7.94} = 0.17$$

则校正装置的传递函数为

$$W_c = \frac{Ts+1}{\gamma_i Ts+1} = \frac{s/1.35+1}{s/0.17+1} \quad (4.8)$$

滞后校正后单自由度机械臂的伯德图如图 4. 6 所示。

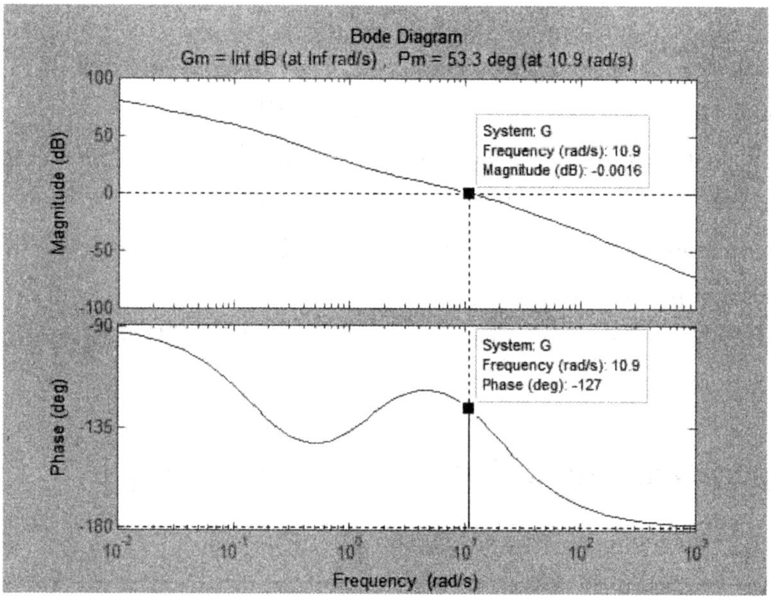

Figure 4.6 Bode Diagram of POFR – Arm after Compensation

图 4.6 滞后校正后单自由度机械臂的伯德图

The step response curve of the system after compensation is shown as Figure 4.7.

滞后校正后系统的阶跃响应曲线如图 4.7 所示。

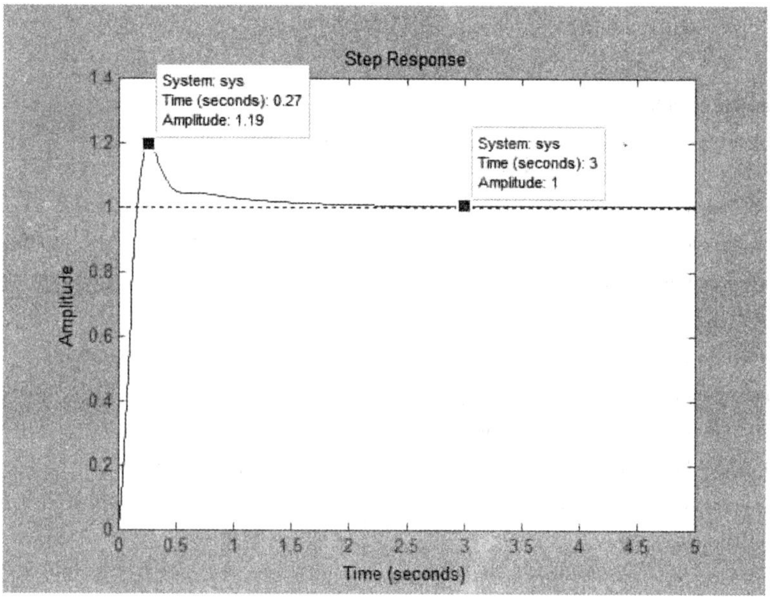

Figure 4.7 Step Response Curve of POFR – Arm after Compensation

图 4.7 系统校正后系统阶跃响应曲线

Obviously:

The phase margin after compensation is: $\gamma = 53.3063°$;

The break frequency after compensation is: $\omega_c = 10.9265$.

After compensation, the system satisfies the requirements of performance index.

MATLAB programming code is shown in Appendix 4.3.

可见：

校正后系统的相角裕度为 $\gamma = 53.3063°$；

校正后系统的截止频率 $\omega_c = 10.9265$。

校正后的系统满足性能指标要求。

MATLAB 程序实现代码见附录 4.3。

4.3 Experiment Instruction

4.3 实验指导书

姓名（Name）:	院系（School）:
专业（Major）:	学号（Student ID）:
起至日期： 年 月 日 至 年 月 日 from: Y M D to Y M D	
实验题目： 频率法校正实验 Experiment Title: Frequency Compensatior Experiment	
Ⅰ. Known Transfer Function and Parameter One freedom robot arm is the controlled plant, The transfer function is $G(s)=\frac{4.587}{s(0.054s+1)}$; **Ⅱ. Experimental Requirements:** 1. Draw the Bode diagram with MATLAB; 2. Write experimental step of lead and lag compensator design; 3. For the same one freedom robot arm, with requirements after compensation: system steady - state speed error constant $K_v \geqslant 5s^{-1}$, phase margin $\gamma(\omega_c) \geqslant 80°$, determine the transfer function of lead and lag compensator; 4. Draw the step response curve after Compensation. **Ⅲ. Experiment Content** 1. According to the transfer function of controlled object, wirte MATLAB code of Bode diagram and test it; 2. According to the requirement , design transfer function of lead and lag compensation; 3. Connect computer and controlled object, do the experiment, and save the Bode diagram and step response curve after compensation. **Ⅳ. Print the Output Result** **Ⅴ. Analyze the Result**	**一、已知技术参数与传递函数** 实验单自由度机械臂为被控对象，传递函数为： $G(s)=\frac{4.587}{s(0.054s+1)}$ **二、实验要求** 1. 用 MATLAB 画出系统的伯德图； 2. 写出超前、滞后校正的实验步骤； 3. 同样针对单自由度机械臂系统，要求校正后的系统稳态速度系数 $K_v \geqslant 5s^{-1}$，相位裕度 $\gamma(\omega_c) \geqslant 80°$，确定超前、滞后校正装置传递函数； 4. 画出校正后的系统响应曲线。 **三、实验内容** 1. 根据被控对象的传递函数，运用 MATLAB 编写伯德图绘制程序并进行实验； 2. 按照要求设计超前、滞后校正传递函数； 3. 连接计算机与被控对象，进行校正实验，并保存校正后系统的伯德图与单位阶跃响应图。 **四、打印实验输出结果** **五、实验结果分析**

4.4　Requirements of the Experiment Report

1) Write every step of lead and lag frequency compensation method.

2) Know the corresponding impact of lead and lag compensation on system.

3) Connect computer with robot arm, doing the physical experiment.

4) Compare with root locus compensation method, and sum up the same and different aspect.

5) Complete the experiment report.

4.4　实验报告要求

1）写出频率法超前、滞后校正的具体计算步骤；

2）掌握超前、滞后校正对系统性能分别有什么影响；

3）计算机连通机械臂，进行实物设计实验；

4）与根轨迹法校正法做对比，总结两种方法的异同点；

5）完成实验报告。

4.5　Experiment Notes

1) Preview before the experiment.

2) Be care of laboratory equipment.

4.5　实验注意事项

1）做实验之前要预习。

2）爱护实验设备。

附录 4.1　画出伯德图的 MATLAB 程序

Appendix 4.1　MATLAB code of Drawing Bode diagram

```
num = [100];
den = [0.054 1 0];
f = tf(num,den)
bode(f)
S = allmargin(f)
```

附录 4.2　超前校正的 MATLAB 程序

Appendix4.2　MATLAB code of Lead compensation

```
num = [100/31.14 100];
den = [0.054/113.69 (1/113.69 +0.054) 1 0];
f = tf(num,den)
bode(f)
S = allmargin(f)
```

附录 4.3　滞后校正的 MATLAB 程序

Appendix4.3　MATLAB code of Lead compensation

```
num = [100/1.35 100];
den = [0.054/0.17 (1/0.17 +0.054) 1 0];
f = tf(num,den)
bode(f)
S = allmargin(f)
```

Chapter 5 Pole Assignment Experiment

第 5 章 极点配置实验

The dynamic performance of the control system is defined by the location of the closed loop poles in *s* plane, so the system dynamic performance requirements should be realized by designing a pair of expected closed loop poles (dominant pole). ***n*** - th, there are ***n*** state variables in. If we see them as the feedback signal of the system, under certain precondition, the arbitrary pole assignment of the system can be realized. The precondition is that the system is controllable, namely, satisfies the system controllability criterion.

由于控制系统的动态性能由它的闭环极点在 s 平面上的位置所确定，因而对系统动态性能的要求可以通过设计一对希望的闭环极点（主导极点）来实现。一个 ***n*** 阶系统有 ***n*** 个状态变量，如果把它们作为系统的反馈信号，在满足一定的前提条件时，就能实现对系统极点的任意配置。所谓的前提条件是系统能控，即满足系统可控性判据。

5.1 Pole Assignment Principle Introduction

5.1 极点配置原理简介

Using the state space method, we can obtain the block diagram, state equation and output equation shown as Figare 5.1 and formula (5.1)

$$\begin{cases} \dot{x} = Ax + Bu \\ y = Cx \end{cases} \tag{5.1}$$

运用状态空间法获得的控制系统的框图、状态方程与输出方程如图 5.1 及式（5.1）所示：

$$\begin{cases} \dot{x} = Ax + Bu \\ y = Cx \end{cases} \tag{5.1}$$

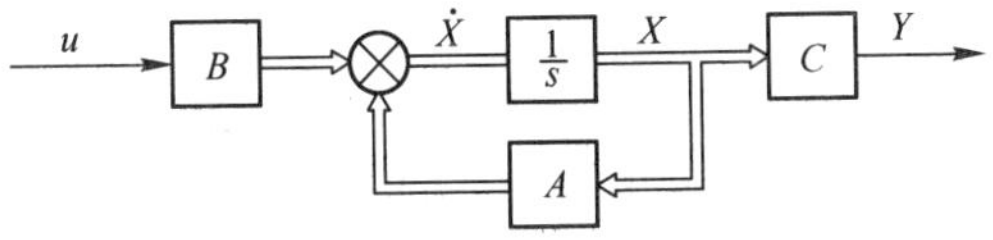

Figure 5.1 Block Diagram of Control System of State Space Model

图 5.1 状态空间模型形式控制系统框图

Where ***x*** denotes the state vector, ***y*** is the output vector, ***u*** is input vector; ***A***, ***B***

其中：*x* 为状态向量，*y* 为输出向量，***u*** 为输入向量；*A*、*B*、*C* 均为与系

and **C** are coefficient matrix related with the system structure and parameters.

By using the Laplace transform for the state space, the matrix of system transfer function will be obtained:

$$G(s) = \frac{Y(s)}{X(s)} = C(SI - A) - 1B$$
$$= C\frac{\mathrm{adj}(sI - A)}{\det(sI - A)}B \qquad (5.2)$$

The characteristic equation of the system is:

$$\det(sI - A) = |sI - A| = 0 \qquad (5.3)$$

The roots of equation (5.3) is the characteristic root of system ($s_1, s_2, \cdots, s_n$), the characteristic roots can reflect the stability of the system and the main dynamic performance. A series correction method can change pole position to improve system performance, but it may not be able to make the system poles in the ideal position (that is, to realize the optimal control). And because of increasing the order of the system (series correction link itself has more than one order and open loop poles), the complexity of the control for system increases.

Adopting the state feedback method means regarding all ***n*** states in the system as feedback variables, feeding back to the input side of the system, and changing the state of the system through the input variable ***u***, the system block diagram is shown as Figure 5.2.

统的结构和参数有关的系数矩阵。

如果对该状态空间模拟运用 Laplace 变换，可以求出系统的传递函数阵为：

$$G(s) = \frac{Y(s)}{X(s)} = C(SI - A) - 1B$$
$$= C\frac{\mathrm{adj}(sI - A)}{\det(sI - A)}B \qquad (5.2)$$

即系统的特征方程为：

$$\det(sI - A) = |sI - A| = 0 \qquad (5.3)$$

式（5.3）的根就是系统的特征根（$s_1, s_2, \cdots s_n$），这些特征根可以反映出系统的稳定性和主要的动态性能。采用串联校正的方法可以使极点位置发生变化以改善系统性能，但不一定能使系统极点处于理想的位置（即实现最优控制）；而且因为增加了系统的阶数（串联校正环节本身具有一阶及以上的开环极点），系统控制的复杂度增大了。

如果采用状态反馈的方式，则意味着将系统中所有 ***n*** 个状态均作为反馈变量，反馈到系统的输入侧，通过输入变量 ***u*** 来改变系统的状态，系统的框图变为图 5.2 所示。

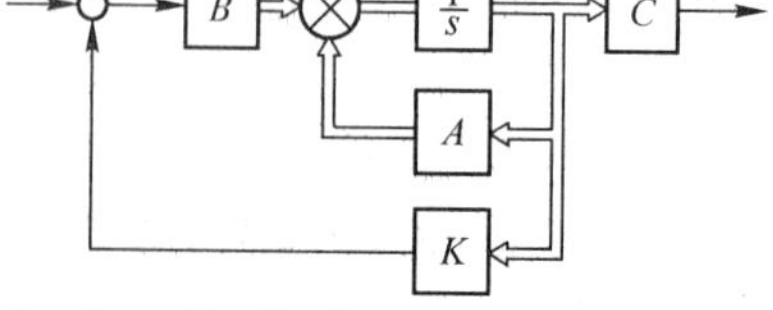

Figure 5.2 State Feedback Control System Block Diagram

图 5.2 状态反馈控制系统框图

The state space model of state feedback control system is:

$$\begin{cases} \dot{x} = Ax + Bu = (A - BK)x + Br \\ y = Cx \end{cases} \quad (5.4)$$

The $\boldsymbol{r}$ denotes the actual input vector; $\boldsymbol{K}$ denotes the state feedback coefficient matrix.

对应于状态反馈的控制系统的状态空间模型为:

$$\begin{cases} \dot{x} = Ax + Bu = (A - BK)x + Br \\ y = Cx \end{cases} \quad (5.4)$$

其中 $\boldsymbol{r}$ 为实际输入向量; $\boldsymbol{K}$ 为状态反馈系数矩阵。

The system characteristic equation is as following:

$$|sI - (A - BK)| = 0 \quad (5.5)$$

Obviously, selecting the appropriate $\boldsymbol{K}$ (where $K = [k_1, k_2, \ldots, k_n]$) can configure the expected characteristic root for control system. There are still $\boldsymbol{n}$ poles after reconfiguration (that is, the state feedback doesn't increase the order of system).

此时系统的特征方程变为:

$$|sI - (A - BK)| = 0 \quad (5.5)$$

显然,选择合适的 $\boldsymbol{K}$ 值($K = [k_1, k_2, \cdots, k_n]$),就可以配置控制系统所希望的特征根。重新配置后的极点仍然只有 n 个(即状态反馈不增加系统的阶次)。

5.2 Analysis of the POFR - Arm Poles Assignment to Improve System Performance

5.2 单自由度机械臂极点配置对改善控制系统性能的分析

5.2.1 State Space Model of the POFR - Arm System

Given that the POFR - Arm system is a second - order system, its state feedback closed loop system block diagram is shown as Figure 5.3.

5.2.1 单自由度机械臂系统的状态空间模型

已知单自由度机械臂系统为一个二阶系统,其状态反馈闭环系统的框图如图 5.3 所示。

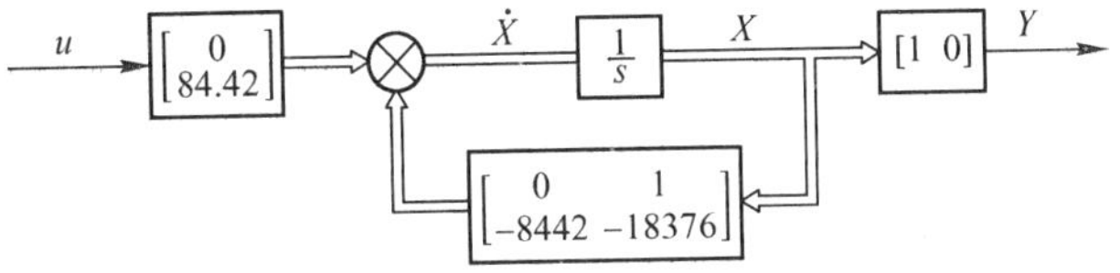

Fig 5.3 Block Diagram of POFR - Arm State Feedback Control System

图 5.3 单自由度机械臂状态反馈控制系统框图

From the block diagram, we can get the system state equation:

$$\begin{cases}\dot{X}=\begin{pmatrix}0 & 1\\ -84.42 & -18.376\end{pmatrix}X+\begin{pmatrix}0\\ 84.42\end{pmatrix}u\\ y=[1\quad 0]X\end{cases}\tag{5.6}$$

由方框图可得系统状态方程为：

$$\begin{cases}\dot{X}=\begin{pmatrix}0 & 1\\ -84.42 & -18.376\end{pmatrix}X+\begin{pmatrix}0\\ 84.42\end{pmatrix}u\\ y=[1\quad 0]X\end{cases}\tag{5.6}$$

The unit step response curve is shown as Figure 5.4.

单位阶跃响应曲线如图 5.4 所示。

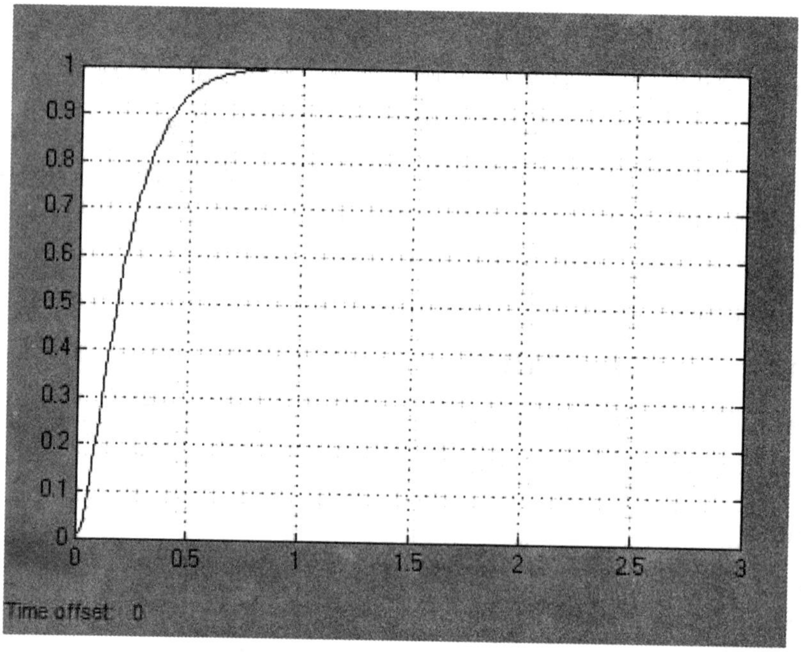

Figure 5.4 Unit Step Response Curve of POFR – Arm System before Pole Assignment

图 5.4 极点配置前的单自由度系统阶跃响应曲线

5.2.2 Check the Condition of Controllability and Pole Assignment

5.2.2 检查系统的能控性与极点配置条件

According to the controllability criterion:

$$\operatorname{rank}[B\quad AB]$$
$$=\operatorname{rank}\begin{bmatrix}0 & 1\\ -84.42 & -18.376\end{bmatrix}$$
$$=2$$

根据能控性判据可得：

$$\operatorname{rank}[B\quad AB]$$
$$=\operatorname{rank}\begin{bmatrix}0 & 1\\ -84.42 & -18.376\end{bmatrix}$$
$$=2$$

The system meet the conditions of controllability, so it should be arbitrary poles assignment.

满足能控性条件，所以系统可以任意配置极点。

Assuming that we need to satisfy the following performance index: percent overshoot $P.O.\% \leqslant 5.4\%$, the peak time $t_p \leqslant 0.5$ s, we can get the expected closed - loop poles:

假定我们需要满足以下的性能指标：超调量 $P.O.\% \leqslant 5.4\%$，峰值时间 $t_p \leqslant 0.5$ s，可以求得希望的闭环极点：

$$S_{1,2} = -\xi\omega_n \pm j\omega_n\sqrt{1-\xi^2} = -6.656 \pm 11.160$$

The expected closed - loop characteristic polynomialis:

希望的闭环特征多项式为：

$$T(s) = s^2 + 13.312s + 168.84$$

Determine the state feedback coefficient k_1 and k_2:

确定状态反馈系数 k_1 和 k_2：

The state feedback of the characteristic equation is as following:

引入状态反馈后的特征方程为：

$$f = s^2 + (18.376 + 84.42k_2)s + 84.42 + 84.42k_1$$

We will get the solution

解得：

$$k_1 = 1, k_2 = -0.06$$

The system block diagram with state feedback is shown as Figure 5.5.

画出引入状态反馈后的系统框图如图5.5所示。

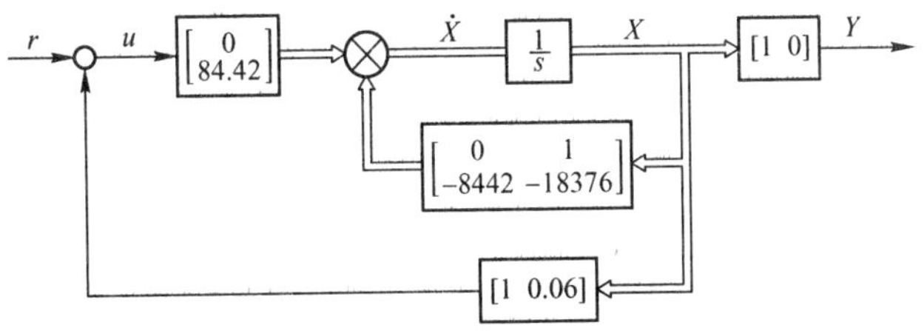

Figure 5.5 System Block Diagram after State Feedback in POFR - Arm

图5.5 引入状态反馈后的单自由度机械臂系统框图

The unit step response curve after pole assignment is shown in Figure 5.6.

极点配置后的单位阶跃响应曲线如图5.6所示。

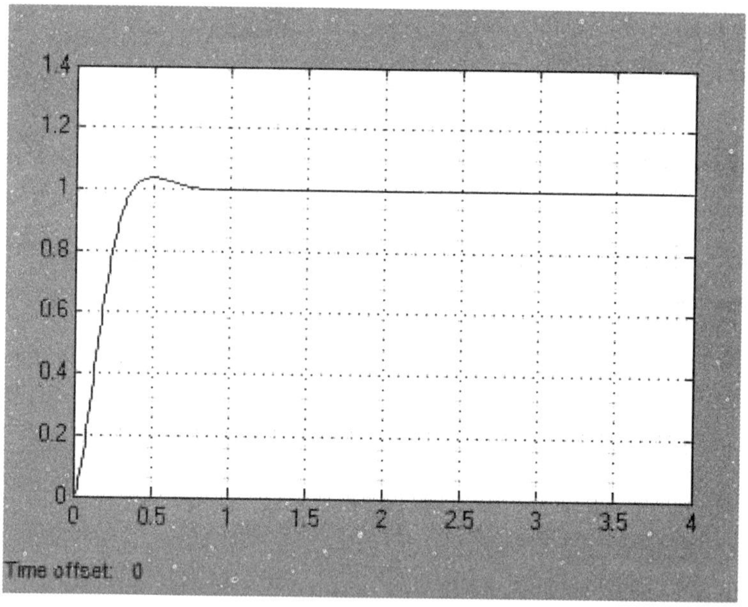

Figure 5.6 The Unit Step Response Curve after Pole Assignment

图 5.6 极点配置后系统的阶跃响应曲线

From Figure 5.6, the percent overshoot $P.O.\% = \frac{1.0369-1}{1} \times 100\% = 3.6\%$, the peak time $t_p \leqslant 0.5s$. So the system satisfies the performance requirements.

根据图 5.6 可得出，系统超调 $P.O.\% = \frac{1.0369-1}{1} \times 100\% = 3.6\%$；峰值时间 $t_p \leqslant 0.5s$。所以系统满足性能指标要求。

5.3 Experiment Instruction

5.3 实验指导书

姓名（Name）：					院系（School）：				
专业（Major）：					学号（Student ID）：				
起至日期： from:	年 Y	月 M	日 D	至 to	年 Y	月 M	日 D		
实验题目：极点配置实验 Experiment Title: Pole Assignment Experiment									

Ⅰ. The Technical Parameters and Transfer Function

Controlled object is robot arm, the transfer function is $G(s) = \frac{4.587}{s(0.054s+1)}$; State equation:

$$\begin{cases} \dot{X} = \begin{bmatrix} 0 & 1 \\ -84.42 & -18.376 \end{bmatrix} X + \begin{bmatrix} 0 \\ 84.42 \end{bmatrix} u \\ y = [1 \quad 0] X \end{cases}$$

Ⅱ. Experimental Requirements

1. Draw the unit step response curve of robot arm before

一、已知技术参数与传递函数

实验单自由度机械臂为被控对象，传递函数为：$G(s) = \frac{4.587}{s(0.054s+1)}$；状态方程为：

$$\begin{cases} X = \begin{bmatrix} 0 & 1 \\ -84.42 & -18.376 \end{bmatrix} X + \begin{bmatrix} 0 \\ 84.42 \end{bmatrix} u \\ y = [1 \quad 0] X \end{cases}$$

二、实验要求

1. 用 MATLAB 画出极点配置前的单自由度机械臂

（续）

<table>
<tr><td>姓名（Name）：
专业（Major）：
起至日期： 年 月 日 至 年 月 日
from： Y M D to Y M D</td><td>院系（School）：
学号（Student ID）：</td></tr>
<tr><td colspan="2">实验题目： 极点配置实验
Experiment Title：Pole Assignment Experiment</td></tr>
<tr><td>pole assignment using MATLAB；
2. Set index and get the state feedback matrix K；
3. Obtain the unit matrix and carry on pole assignment；

Ⅲ. The Content of Experiment
1. According to the transfer function, write the program of response for unit step input. And give out the response curve；
2. According to the requirement, write the MATLAB program of feedback matrix K；
3. Connect the computer and robot arm, to do the real plant experiment, and compare with simulation. Save the results of response curves；
4. Print the results；
5. Analysis the results.</td><td>系统的单位阶跃响应曲线；
2. 设定指标，以求得状态反馈矩阵 K；
3. 针对求得的单位矩阵，进行极点配置；
4. 画出极点配置后的单自由度机械臂的单位阶跃响应曲线。
三、实验内容
1. 根据被控对象的传递函数，运用 MATLAB 编写单位阶跃响应程序并进行实验；
2. 按照要求，编写求得系统的状态转移矩阵 K 的 MATLAB 程序；
3. 连接计算机与被控对象，进行对比实验，并保存极点配置前后系统的单位阶跃响应图；
4. 打印实验输出结果；
5. 实验结果分析。</td></tr>
</table>

5.4 Requirements of the Experiment Report

1）Write the steps of pole assignment；

2）Grasp the effect that pole assignment has on system performance；

3）Connect one freedom robot arm with computer, then carry on the physical design experiment；

4）Change the data of pole, and compae the performance change of the system unit step response；

5）Complete experiment report.

5.4 实验报告要求

1）写出极点配置的具体计算步骤；

2）掌握极点配置对系统性能有什么影响；

3）电脑连通单自由度机械臂，进行实物设计实验；

4）改变极点数据，比较系统单位阶跃响应的性能变化；

5）完成实验报告。

5.5 Experiment Notes

1）Preview before doing experiment；

2）Take care of the experimental equipments.

5.5 实验注意事项

1）做实验之前要预习；

2）爱护实验设备。

Chapter 6 Design Experiment of PID Controller

In engineering field, the most widely used control laws in regulator are proportional, integral and differential control, hereinafter referred to as PID control.

Adventure of PID controller has been nearly 70 years history, as for its simple structure, good stability, reliable operation, and easy to adjust, the PID method becomes one of the main technology in industry. When the structure and parameters of plant are not clear, or there is no precise mathematical model to describe the system, even other control theory and technology are hard to use, so the structure and parameters of system controller have to be confirmed by experience and on - site adjusting, the application of PID technology is the most suitable in such situation.

When we don't fully understand asystem and the plant, or can't obtain system parameters through effective measurement method, the most suitable algorithm is to design a PID controller.

PID controller includes PI, PD and PID controller respectively. According to the error of system, by means of the proportion, integral, differential, we can calculate the control law to control a real system. Block diagram of POFR - Arm with PID controller is shown as Figure. 6. 1.

第6章 PID控制器设计实验

在工程实际中，应用最为广泛的调节器控制规律为比例、积分、微分控制，简称PID控制。

PID控制器问世至今已有近70年历史，它以其结构简单、稳定性好、工作可靠、调整方便而成为工业控制的主要技术之一。当被控对象的结构和参数不能完全掌握，或得不到精确的数学模型，控制理论的其他技术难以采用时，则系统控制器的结构和参数不得不依靠经验和现场调试来确定，这时应用PID控制技术最为方便。

即当我们不完全了解一个系统和被控对象，或不能通过有效的测量手段来获得系统参数时，最适合使用PID技术来进行系统控制。

PID控制，分别有PI、PD和PID控制。PID控制器就是根据系统的误差，利用比例、积分、微分计算出控制律进行控制的。机械臂PID控制系统的框图如图6.1所示。

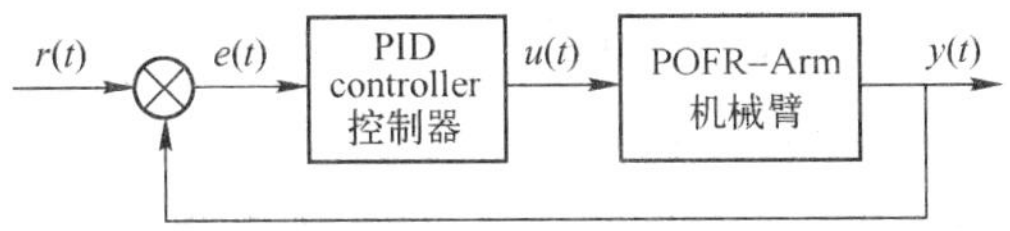

Figure 6. 1 Block Diagram of POFR - Arm + PID Controller

图 6. 1 单自由度机械臂 + PID 控制器的闭环控制系统框图

Proportional control (P): It adjusts the error of system with proportion. Once the system has deviation, proportional unit begins the adjustment immediate to reduce the deviation. Increase the proportional parameter P can improve the effect of adjustment and reduce the error. But overlarge proportion P will reduce the stability, and even make the system unstable.

比例调节（P）：对系统的偏差进行比例调节。系统一旦出现了偏差，比例调节立即产生调节作用以减少偏差。比例作用大时，可以加快调节，减少误差，但是过大的比例，使系统的稳定性下降，甚至造成系统的不稳定。

Integral control (I): It will eliminate the steady - state error of the system. Because of the system error, the integral control adjusts it until there is no differential, and then integral control stops, the output of integral control is constant. The strength of the integral action depends on integrating gain T_I. The smaller the T_I, the stronger the integral action. Wndvice versa The integral control can make the system stability and dynamic response slow. Integral action is often combined with the other two regulating laws, formed PI regulator or PID regulator.

积分调节（I）：可以使系统消除稳态误差。当有误差时，积分调节就一直进行直至系统误差为零，此后积分调节停止，积分调节单元的输出为常值。积分作用的强弱取决于积分环节增益参数 T_I，T_I 越小，积分作用就越强。反之，T_I 大则积分作用弱，加入积分调节会使系统稳定性下降，动态响应变慢。积分作用常与另两种调节规律结合，组成 PI 调节器或 PID 调节器。

Differential control (D): Differential effect reflects the rate of change of deviation signal, it has characteristic of prediction, and will control the system forecasting according to the trend of deviation change. It can produce the advancing control action to eliminate the deviation when deviation has not formed yet.

微分调节（D）：微分作用反映系统偏差信号的变化率，具有超前性，能根据偏差变化的趋势，产生超前的控制作用。在偏差还没有形成之前，已被微分调节作用消除。

Differential controller can improve the 8ystem dynamic performance. Appropriate differential gain can reduce the overshoot, settling time. But differential controller will amplify the noise in the system, so it is bad for anti - jamming when put on a strong differential regulation. As the differential indicates the rate of vary, so when the input does not change, the output of differential controller will be zero. Differential cannot be used alone, it needs combined with two other regulation laws, formed PD or PID controller.

微分调节可以改善系统的动态性能。在微分增益选择合适的情况下，可以减少超调和调节时间。微分作用对噪声干扰有放大作用，因此过强的微分调节，对系统抗干扰不利。此外，微分反应的是变化率，而当输入没有变化时，微分作用输出为零。微分作用不能单独使用，需要与另外两种调节规律结合，组成 PD 或 PID 控制器。

6.1 Analysis on Mathematical Model of PID Controller

6.1 PID 控制器的数学模型分析

Classical PID control law for continuous time is shown as formula (6.1):

$$u(t) = K_p\left\{e(t) + \frac{1}{T_I}\int_0^t e(\tau)\mathrm{d}\tau + T_d\frac{\mathrm{d}e(t)}{\mathrm{d}t}\right\} \tag{6.1}$$

Where $u(t)$ denotes the controlled quantity, $e(t)$ denotes error, K_P, T_I, T_D denote the proportional gain, integral time constant and differential time constant respectively. In compater control system, The digital form of the PID controller is used, it is shown as (6.2):

$$u(t) = K_p\left\{e(k) + \frac{1}{T_I}\sum_{i=0}^{k} e(i)T + T_d\frac{e(k) - e(k-1)}{T}\right\} \tag{6.2}$$

where, T is sampling period.

Formula (6.2) is the most primitive form of the digital PID controller, each time the computing needs $k+5$ times of multipli-

经典 PID 连续时间控制律如式 (6.1) 所示：

$$u(t) = K_P\left\{e(t) + \frac{1}{T_I}\int_0^t e(\tau)\mathrm{d}\tau + T_d\frac{\mathrm{d}e(t)}{\mathrm{d}t}\right\} \tag{6.1}$$

其中 $u(t)$ 为控制量，$e(t)$ 为误差，K_P、T_I、T_D 分别为比例增益、积分时间常数和微分时间常数。在计算机控制系统中使用 PID 控制器数字形式，如式 (6.2) 所示：

$$u(k) = K_P\left\{e(k) + \frac{1}{T_i}\sum_{i=0}^{k} e(i)T + T_d\frac{e(k) - e(k-1)}{T}\right\} \tag{6.2}$$

其中 T 为采样周期。

式 (6.2) 为数字 PID 控制器最原始的形式，每次计算需要 $k+5$ 次乘法运算，$k+1$ 次加法运算，而且需要保存 $k+1$ 个

cation, $k+1$ times of addition and needs to save the value of the variable $k+1$ times. With the system running, k value increases, operation becomes more slowly, and the storage space grows linearly. In order to reduce computational complexity, increase the operation speed and reduce the influence of operation on the system real - time performance, the formula needs deformation:

$$u(k) = K_1 e(k) + K_2 e(k-1) + K_3 \sum_{i=1}^{k-2} e(i) \qquad (6.3)$$

where,

$$K_1 = K_p + \frac{K_p T}{T_I} + \frac{K_p T_d}{T}$$
$$K_2 = \frac{K_P T}{T_I} - \frac{K_p T_d}{T} \qquad (6.4)$$
$$K_3 = \frac{K_P T}{T_I}$$

Each factor in formula (6.4) is calculated during initialization, each time only needs three multiplication and two addition operations, and does not change with the increasing of k, thus ensuring the consistency of computational complexity; Each sample only needs to save the two variables $e(k-1)$, $\sum(k-2)$, and does not change along with the increasing of k value.

Therefore, in PID control experiment, there are three parameters for design, K_p, T_I, T_d. And, by adjusting the three parameters, students can understand the PID system principle and the actual control effect performance.

变量的值，随着系统的运行 k 值不断增大，运算也更为缓慢，而且占用的存储空间也线性增长。为了减小运算量，提高运算速度，减小运算对系统实时性的影响，对公式变形得到：

$$u(k) = K_1 e(k) + K_2 e(k-1) + K_3 \sum_{i=1}^{k-2} \cdot e(i) \qquad (6.3)$$

其中，

$$K_1 = K_p + \frac{K_p T}{T_I} + \frac{K_p T_d}{T},$$
$$K_2 = \frac{K_P T}{T_I} - \frac{K_p T_d}{T} \qquad (6.4)$$
$$K_3 = \frac{K_P T}{T_I}$$

式（6.4）中的各个系数可以在初始化时计算一次，每次只需要 3 次乘法和 2 次加法运算，而且不随 k 的增加而变化，保证了计算复杂度的一致性；每次采样只需要保存 $e(k-1)$、$\sum(k-2)$ 两个变量，而且不随 k 值的增加而变化。

因此，在 PID 控制实验中，可以一共设计 3 个参数，分别为 K_p，T_I，T_d。通过调节这三个参数，可以理解 PID 调节系统性能原理及实际控制效果。

6.2 Design Experiment of PID Controller

6.2 PID 控制器设计实验

6.2.1 Proportional (P) Control Experiment

6.2.1 比例（P）控制实验

When using P controller in POFR - Arm, the function of controller is equivalent to a proportional gain K_P, the mathematical model of the system is shown in Figure 6.2.

对便携式单自由度机械臂使用 P 控制进行调节时，控制器的作用相当于一个比例增益 K_p，此时，系统的数学模型如图 6.2 所示。

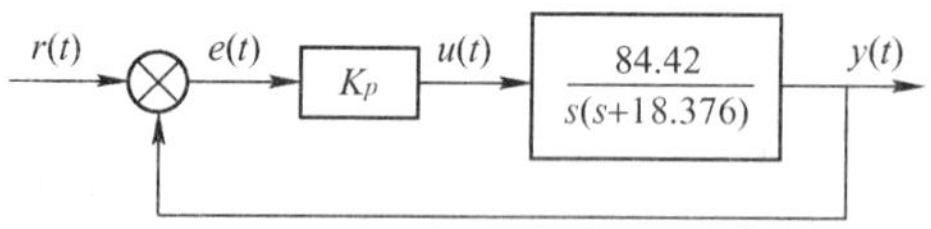

Figure 6.2 Closed Loop Block Diagram of POFR - Arm + P Controller

图 6.2 单自由度机械臂 + P 控制器的闭环控制系统框图

From Figure 6.2, the system is a standard second - order system, the open loop transfer function is:

$$G(s) = \frac{84.42K_p}{s(s+18.376)} \quad (6.5)$$

The closed - loop transfer function is:

$$G(s) = \frac{84.42K_p}{s^2+18.376s+84.42K_p} \quad (6.6)$$

According to the automatic control theory in the time domain analysis method of second order system, we can analyze, when $K_P < 1$, the system has two negative real roots, therefore, the system is stable and doesn't have overshoot. When $K_P > 1$, the system has two conjugate complex roots, and the real part is -9.188, namely the two poles of the system are

$$s_{1,2} = -9.188 \pm j\sqrt{84.42K_p - 84.42} \quad (6.7)$$

从图 6.2 得出，该系统是一个标准的二阶控制系统，系统开环传递函数变为：

$$G(s) = \frac{84.42K_p}{s(s+18.376)} \quad (6.5)$$

闭环传递函数为：

$$G(s) = \frac{84.42K_p}{s^2+18.376s+84.42K_p} \quad (6.6)$$

根据自动控制理论中二阶系统时域分析方法可以分析出来，当 $K_p < 1$ 时，此时系统存在两个负数实根，因此，系统稳定且无超调。当 $K_p > 1$ 时，系统将出现两个共轭复根，且根的实部为 -9.188，即系统的两个极点为

$$s_{1,2} = -9.188 \pm j\sqrt{84.42K_p - 84.42} \quad (6.7)$$

It can be calculated, if K_P increases continually, the real part is keep unchanged, imaginary part keep increase continually too; it will lead to large overshoot.

可以算出如果继续增大 K_p，极点实部不变，虚部继续增大，则导致超调量会随之增加。

Using MATLAB for simulation experiments, and physical experiment of POFR - Arm system respectively, we can get the results as shown in Figure 6.3 and Figure 6.4, three curves in the graph are response curves with parameter K_P 1, 4, 8 respectively.

运用 MATLAB 进行仿真实验，以及单自由度机械臂系统的实物实验可以得到实验结果如图 6.3、图 6.4 所示，图形中三条曲线分别是 K_p 参数为 1，4，8 时的响应曲线。

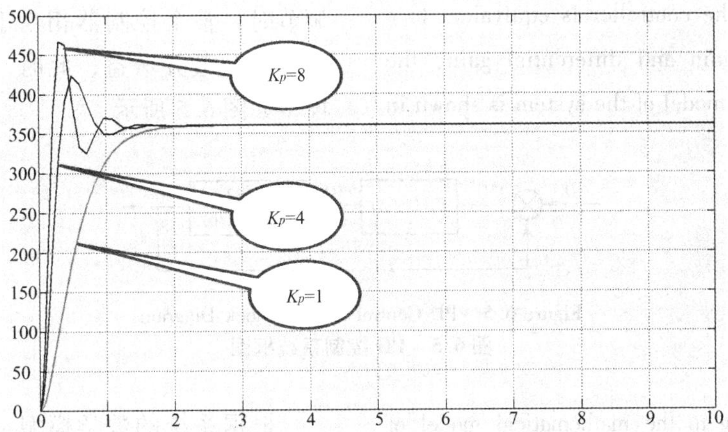

Figure 6.3 MATLAB Simulation Curve for Proportional Control

图 6.3 比例控制的 MATLAB 仿真曲线

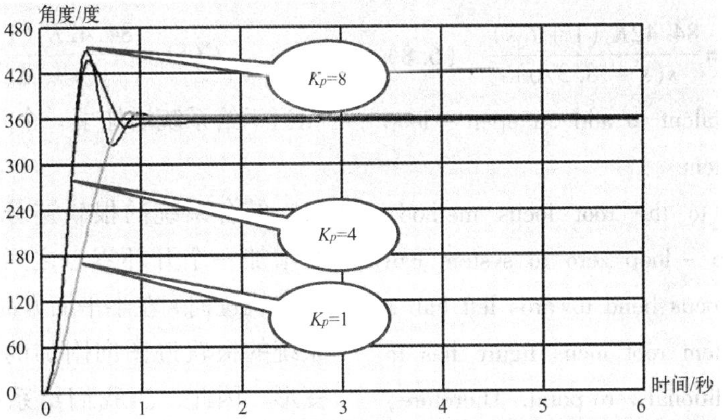

Figure 6.4 Experimental Curves of POFR - Arm for Proportional Control

图 6.4 比例控制的单自由度机械臂实物实验曲线

According to Figure 6. 3 and Figure 6. 4, we can confirm that using P control, the increase of the proportionality coefficient parameters K_P can speed up the corresponding rate of system, increasing the overshoot of system. The results of simulation and real experiment results are consistent with the basic theory.

根据图 6. 3 和图 6. 4，可以确认使用 P 控制，随着比例系数 K_p 参数的增大，确实可以加快系统的响应速率，增大系统的超调量。仿真结果和实物实验效果与理论分析基本一致。

6. 2. 2 Proportional Derivative (PD) Control Experiment

6. 2. 2 比例微分（PD）控制实验

When the POFR - Arm is adjusted by PD control, the controller is equivalent to a proportional gain and differential gain, the mathematical model of the system is shown in Figure 6. 5:

对单自由度机械臂使用 PD 控制进行调节时，整个控制器相当于一个比例增益和一个微分增益，此时，系统的数学模型如图 6. 5 所示。

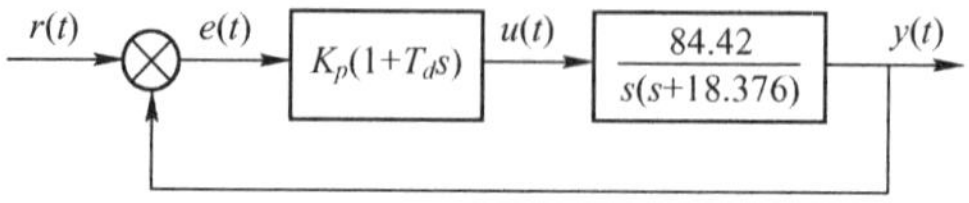

Figure 6. 5 PD Control System Block Diagram

图 6. 5 PD 控制系统框图

According to the mathematical model of the system, it can be found that in POFR - Arm system after adding the PD controller, open - loop transfer function of the system is:

$$G(s) = \frac{84.42K_p(1+T_d s)}{s(s+18.376)} \quad (6.8)$$

It's equivalent to add an open - loop zero for the system.

根据系统的数学模型可以发现，在单自由度机械臂系统中加入 PD 控制器之后，系统的开环传递函数是：

$$G(s) = \frac{84.42K_p(1+T_d s)}{s(s+18.376)} \quad (6.8)$$

相当于给系统增加了一个开环零点。

According to the root locus method, adding an open - loop zero to system can make the root locus bend towards left half s plane, the system root locus figure has to close to the additional zero point. Therefore, when we add a negative open loop zero to system, it can significantly improve the stability of the system.

结合系统的根轨迹分析法，当给系统增加一个开环零点之后，可以使系统的根轨迹向 s 左半平面方向弯曲，也就是系统的根轨迹图向附加的零点方向进行变形。因此，当我们给系统增加一个负的开环零点之后，可以显著改善系统的稳定性。

Moreover, adding the open loop zero is equivalent to plus a closed loop zero, increasing the close - loop zero has impact on system dynamic performance, this is equal to the reduce of the closed - loop system damping, so as to make the system transition process has a tendency to appear overshoot, which accelerate the system dynamic response rate.

而且，增加开环零点也就相当于增加了一个闭环零点，增加闭环零点会对系统的动态性能产生影响，相当于减小闭环系统的阻尼，从而使系统过渡过程有出现超调的趋势，即加快了系统的动态响应速率。

Therefore, through the analysis of the root locus method, we' ll get the conclusion: using PD controller in control system can increase the corresponding speed of dynamic response, and improve the stability of the system also.

因此，通过根轨迹分析方法，我们可以得出结论：使用 PD 控制器对系统进行控制，可以增大系统动态响应的响应速率，也能改善系统的稳定性。

Figure 6. 6 and Figure 6. 7 are the results of the MATLAB simulation and physical experiment of POFR - Arm respectively, the graphical curve PD parameters are $K_p = 4$、$T_d = 10$; $K_p = 4$、$T_d = 20$, respectively.

图 6. 6 与图 6. 7 分别是运用 MATLAB 仿真及单自由度机械臂实物实验的结果图，各个图形中曲线 PD 参数分别为$K_p = 4$、$T_d = 10$；$K_p = 4$、$T_d = 20$。

According to Figure 6. 6 and Figure 6. 7, it will be confirmed that using PD controller can increase the corresponding rate of system, reduce the overshoot of system. If we keep *P* parameters unchanged, and increase *D* parameters, the overshoot is reduced, the rising time become faster. The results of simulation and physical experiment are consistent with the theory analysis.

根据图 6. 6 与图 6. 7，我们可以确认使用 PD 控制器，确实可以增大系统的响应速率，降低系统的超调量。而且在 *P* 参数不变的前提下，随着 *D* 参数的增大，超调量降低越多，响应速率提升越快。仿真结果和实物实验效果与理论分析基本一致。

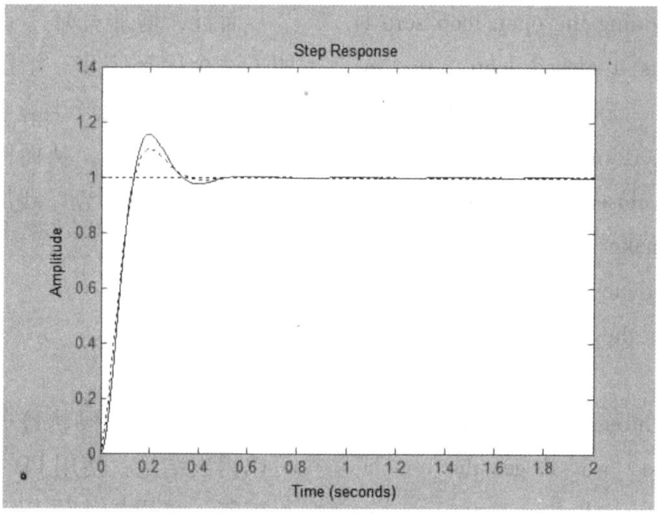

Figure 6.6　MATLAB Simulation Experiment of Unit Step Response with PD Control

图 6.6　PD 控制的 MATLAB 仿真实验单位阶跃响应曲线

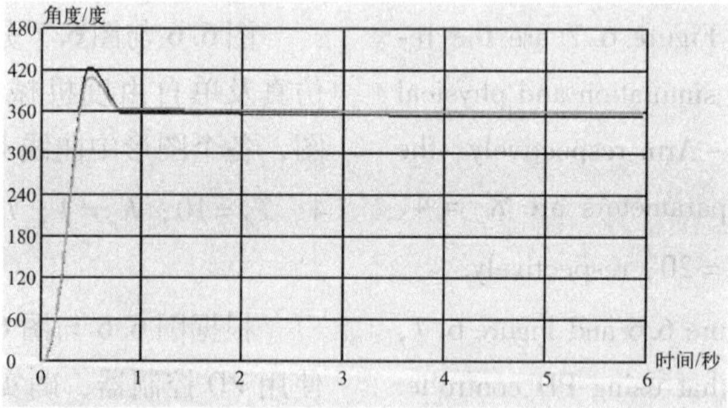

Figure 6.7　Physical (POFR – Arm) Experiment of The Unit Step Response with PD Control

图 6.7　PD 控制的单自由度机械臂实物实验单位阶跃响应曲线

6.2.3 Proportional Integral (PI) Control Experiment

PI Control System block digram is shown in Figure 6.8.

6.2.3 比例积分（PI）控制实验

比例积分（PI）控制系统框图如图 6.8 所示。

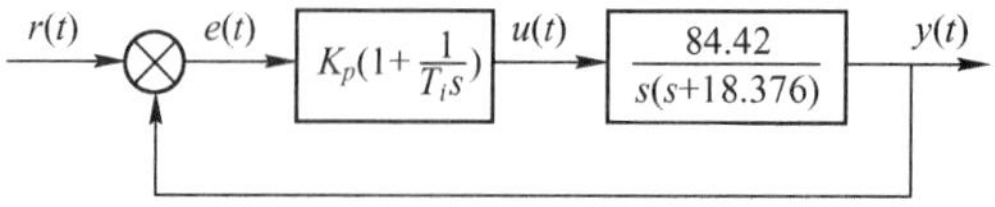

Figure 6.8　PI Control System Block Digram

图 6.8　比例积分控制系统框图

When the controller is PI controller, its transfer function is

$$G_{PI}(s) = K_p\left(1 + \frac{1}{T_i s}\right) \qquad (6.9)$$

Set $K_P = 4$ to constant value, by changing the size of the integral time constant T_i, we can get the system step response curves using MATLAB simulation and physical experiment, respectively. The POFR - Arm curves by experiment are shown in Figure 6.9 and Figure 6.10.

当控制器为 PI 控制器时，其控制器的传递函数为：

$$G_{PI}(s) = K_p\left(1 + \frac{1}{T_i s}\right) \qquad (6.9)$$

设定 $K_p = 4$ 为定值，改变积分时间常数 T_i 大小，可以得到系统的阶跃响应曲线。分别采用 MATLAB 仿真实验和单自由度机械臂实物实验，得到实验曲线如图 6.9 和图 6.10 所示。

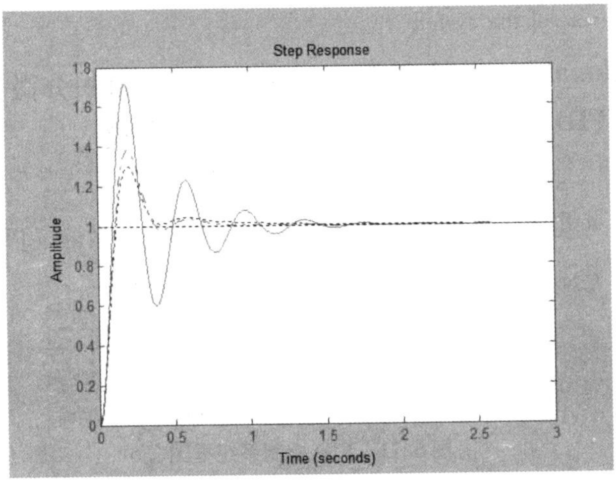

Figure 6.9 Matlab Simulation Experiment of Unit Step Response with PI Controller

图 6.9 PI 控制的 Matlab 仿真实验单位阶跃响应曲线

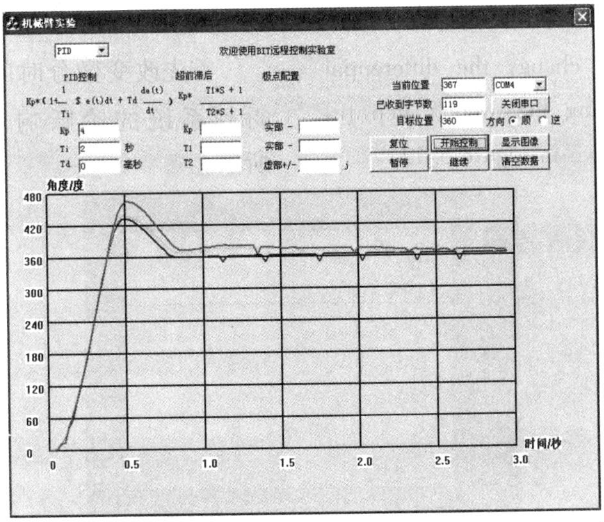

Figure 6.10 Physical Experiment (POFR - Arm) of Unit Step Response with PI Controller

图 6.10 PI 控制的单自由度机械臂实物实验单位阶跃响应曲线

Simulation results show: when $K_P = 4$, with the increasing of value T_i, system overshoot decreases, and the system response speed is slightly slower. In contrast, when the value of T_i reduces gradually, the system overshoot increases, the system response speeds up. The smaller the T_i, the faster the integral, the stronger the integral action, and there are more oscillations. *PI* control can eliminate the steady state error of the system, and increase the precise of the system.

仿真结果表明：当 $K_p = 4$，随着 T_i 值的加大，系统的超调量减小，系统响应速度略微变慢。相反，当 T_i 的值逐渐减小时，系统的超调量增大，系统的响应速度加快。T_i 越小，积分速度越快，积分作用就越强，系统震荡次数较多。*PI* 控制可以消除系统的稳态误差，提高系统的精度。

6.2.4 Proportional Integral Differential (PID) Control Experiment

6.2.4 比例积分微分（PID）控制实验

Block diagram of PID Control system is shown in Figure 6.11.

比例积分微分（PID）控制系统框图如图 6.11 所示。

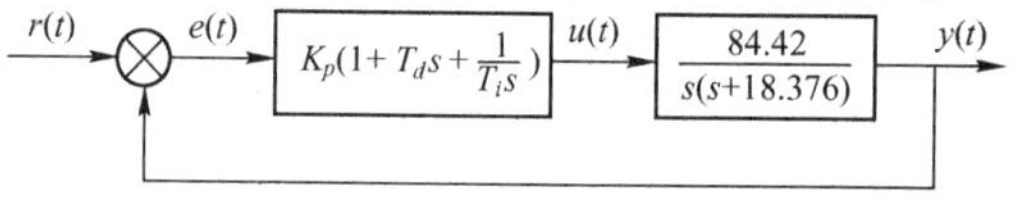

Figure 6.11 Block Diagram of PID Control System
图 6.11 PID 控制系统框图

PID controller transfer function is:

$$G_{PID}(s) = K_p\left(1 + T_d s + \frac{1}{T_i s}\right)$$

PID 控制器的传递函数为：

$$G_{PID}(s) = K_p\left(1 + T_d s + \frac{1}{T_i s}\right)$$

First, we can change the differential time constant, getting the system step response curve as shown in Figure 6.12.

首先改变微分时间常数 T_d 的大小，得到系统的阶跃响应曲线如图 6.12 所示。

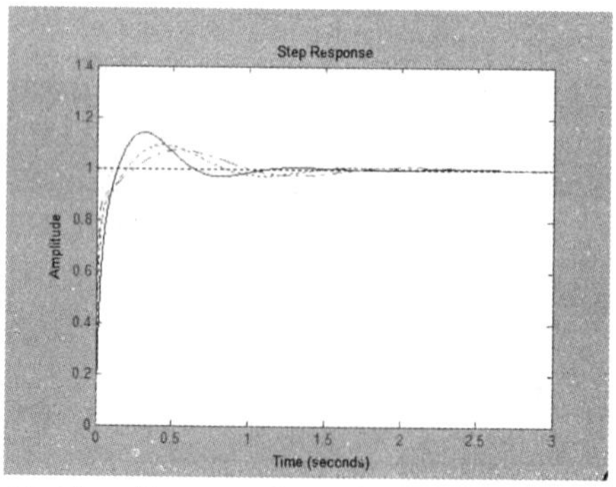

Figure 6.12 Step Response Curve Corresponding to Different T_d
图 6.12 不同 T_d 值对应的阶跃响应曲线

Then select the other appropriate controller parameters, and the PID controller is designed.

然后选定合适的其他控制器参数，设计 PID 控制器。

Based on the above analysis, the choice of $K_P = 4$, $T_i = 0.15$, $T_d = 0.1$ can make the system performance indicators meet the design requirements. After calculation, overshoot $\sigma\% < 20\%$, and $t_p < 0.5\% s$ meet the design requirements. System step response curve is shown as Figure 6.13.

根据上述分析，选择 $K_p = 4$，$T_i = 0.15$，$T_d = 0.1$，可使系统性能指标达到设计要求。经计算，超调量 $\sigma\% < 20\%$，$t_p < 0.5s$ 满足设计要求。系统的阶跃响应曲线如图 6.13 所示。

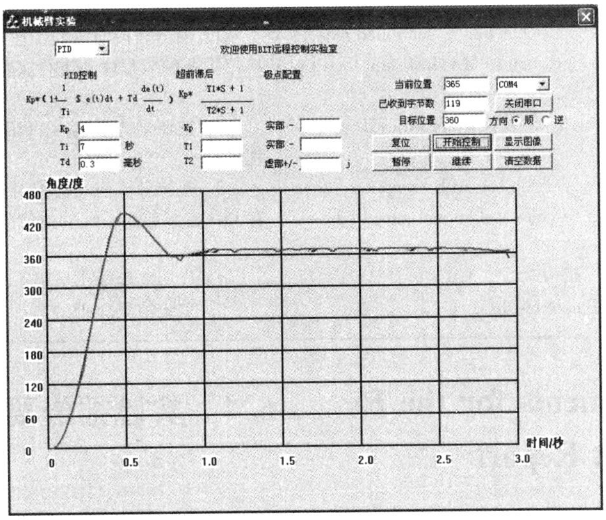

Figure 6.13 System Step Response Curve with PID Controller

图 6.13 加比例积分微分控制器的系统阶跃响应曲线

6.3 Experiment Instruction

6.3 实验指导书

姓名（Name）： 院系（School）：
专业（Major）： 学号（Student ID）：
起至日期： 年 月 日 至 年 月 日
from： Y M D to Y M D

实验题目： PID 实验
Experimental Title：PID Control Experiment

（续）

Ⅰ. The Technical Parameters and Transfer Function Controlled object: Robot arm; Transfer function: $G(s) = \frac{84.42K_p}{s(s+18.376)}$. Ⅱ. Experimental Requirements 1. Draw the unit step response curve of robot arm in MATLAB; 2. Design the PID parameters and connect robot arm with controller toget the closed loop control system; 3. Draw the simulation curve for closed loop control system; 4. Draw the unit step response curve of robot arm with PID controller for closed loop system. Ⅲ. Experimental Content 1. According to the transfer function of controlled object, write the unit step response program in MATLAB and then experiment; 2. As requested, write MALAB program with PID parameter adjustment; 3. Connect the controlled object with computer, then save the figure of unit step response before and after PID control; 4. Print the output result; 5. Analyze the experiment results.	**一、已知技术参数与传递函数** 实验单自由度机械臂为被控对象，传递函数为：$G(s) = \frac{84.42K_p}{s(s+18.376)}$。 **二、实验要求** 1. 用 MATLAB 画出单自由度机械臂系统的单位阶跃响应曲线； 2. 设计 PID 参数，连接控制器与单自由度机械臂，形成闭环控制系统； 3. 画出闭环控制系统的仿真响应曲线； 4. 画出 PID 控制器与单自由度机械臂的闭环系统的实物单位阶跃响应曲线。 **三、实验内容** 1. 根据被控对象的传递函数，用 MATLAB 编写单位阶跃响应程序并进行实验； 2. 按照要求，编写 PID 参数调整的 MATLAB 程序； 3. 连接计算机与被控对象进行对比实验，并保存 PID 控制前后系统的单位阶跃响应图； 4. 打印实验输出结果； 5. 实验结果分析。

6.4 Requirements for the Experiment Report

1) Record the results of simulation and physical experiment;

2) Summarize the characteristics of PID control;

3) Analyze what effect P, I, D may have on system performance;

4) Connect computer with robot arm, and do the physical experiment;

5) Change the pole figures and compare the performance change of system step response;

6) Complete the experiment report.

6.4 实验报告要求

1）记录仿真和上机实验结果；

2）总结 PID 控制的特点；

3）分析 P、I、D 分别对系统性能有什么影响；

4）计算机连通单自由度机械臂，进行实物设计实验；

5）变化极点数据，比较系统单位阶跃响应的性能变化；

6）完成实验报告。

6. 5 Experiment Notes

1) Preview before the experiment.

2) Be care of laboratory equipment.

6. 5 实验注意事项

1) 做实验之前要预习;

2) 爱护实验设备。

Chapter 7 The Linear Quadratic Regulator Optimal Control Algorithm (LQR) Controller Experiment

第7章 线性二次型最优控制算法（LQR）控制器的设计及实现

The most prominent feature of modern control theory is the control object in the form of a state space expression, and it's easy to analyze and design the type of multiple input multiple output system. The linear quadratic regulator optimal control algorithm (LQR) is an important and basic method of modern control theory, under a certain performance index, LQR algorithm could make the system achieve the best control effect, that is to say, using the least control energy to achieve the minimum error. By using the optimal control algorithm, this chapter is mainly to control the swinging rod angle and cart position of the one - stage flexible connection inverted pendulum system at the same time.

现代控制理论的最突出特点就是将控制对象用状态空间表达式的形式表示出来，这样便于对多输入多输出系统进行分析和设计。线性二次型最优控制算法（LQR）是现代控制理论中一种重要的、基本的方法，LQR 算法的目的是在一定的性能指标下，使系统的控制效果最佳，即利用最少的控制能量，来达到最小的状态误差。本章主要利用最优控制算法实现对一级柔性连接倒立摆系统的摆杆角度和小车位置的同时控制。

Design purpose:

Learn how to use the optimal control algorithm for LQR controller design.

设计目的：

学习如何使用最优控制算法设计系统 LQR 控制器。

Design requirements:

Simulation control: Impose a 0.1 step response on the system, the system can quickly adjust to equilibrium state with stabilization time no more than 2 second, and the steady state error is less than 1%.

Actual control: system can be stable within 3s

Performance indicators:

1) The system stable time is less than

设计要求：

仿真控制：对系统施加 0.1 的阶跃响应时，系统能快速调节至平衡状态，调节时间不超过 2s，稳态误差小于 1%。

实际控制：要求系统可以在 3 s 内进行稳定控制。

性能指标为：

1）系统稳定时间小于 3 s；

3 second;

2) The amplitude of θ is less than 0.3 °;

3) The steady - state error is less than 2%.

2）θ 的幅值小于0.3°；

3）稳态误差小于2%。

7.1 The Theoretical Analysis

7.1 理论分析

7.1.1 The LQR Optimal Control Theory

Given the system state equation

$$\dot{X}(t) = AX(t) + BU(t) \qquad (7.1)$$

where

$$U(t) = -KX(t) \qquad (7.2)$$

It is required to determine the optimal control vector $U(t)$ in the matrix, which makes the performance index $\boldsymbol{J}$ to minimum.

$$J = \int_0^{\infty} [X^T(t)QX(t) + U^T(t)RU(t)]dt \qquad (7.3)$$

Q is a positive definite (or positive - semi definite) real symmetric matrix, R is a positive definite real symmetric matrix. The first item to the right of formula (7.3) embodies the requirement for system error, the second reflects the requirement for energy loss of control signal, the relative size of matrix Q and R determine the relative importance of the error and energy loss. Here we assume that control vector $U(t)$ is not constrained. Therefore, based on the LQR performance, the optimal control system design boils down to determine the elements of matrix K. One of the advantages of adopting

7.1.1 线性二次型最优控制原理

已知系统状态方程为

$$\dot{X}(t) = AX(t) + BU(t) \qquad (7.1)$$

其中

$$U(t) = -KX(t) \qquad (7.2)$$

要求确定最优控制向量式（7.2）的矩阵 K，使得性能指标 J 达到最小。

$$J = \int_0^{\infty} [X^T(t)QX(t) + U^T(t)RU(t)]dt \qquad (7.3)$$

式中 Q 是正定（或半正定）实对称矩阵，R 是正定实对称矩阵。式（7.3）右边的第一项体现了对系统误差的要求，第二项体现了对控制信号能量损耗的要求，矩阵 Q 和 R 的相对大小，确定了误差和能量损耗的相对重要性。在此，假设控制向量 $U(t)$ 是不受约束的。因此，基于二次型性能指标的最优控制系统的设计归结为确定矩阵 K 的各元素。采用二次型最优控制方法的一个优点是除了系统不可控的情况外，所设计的系统将是稳定的。

the LQR method is the design of the system will be stable except that the system is uncontrollable.

After the deduction, when the performance of LQR is defined by formula (7.3), the optimal law is linear, and can be given by

$$U(t) = -KX(t) = -R^{-1}B^TPX(t) \tag{7.4}$$

where,

$$K = R^{-1}B^TP \tag{7.5}$$

And matrix P must satisfy the following equation:

$$A^TP + PA - PBR^{-1}B^TP + Q = 0 \tag{7.6}$$

Equation (7.6) is called the degenerate matrix Riccati equation, the steps are as follows:

1) Solve the degenerate matrix Riccati equation (7.6) and get matrix P, if there is a positive definite matrix (some systems may not have positive definite matrix), the system is stable, the matrix A - BK is stable matrix.

2) Generate the matrix P into the equation (7.5), the optimal matrix K is obtained.

MATLAB has a command, *lqr* which gives the continuous time solution of Riccati equation and can determine the optimal feedback gain matrix.

经推导，当二次型最优控制的性能指标由式（7.3）定义时，其最优控制规律是线性的，并由式（7.4）给出。

$$U(t) = -KX(t) = -R^{-1}B^TPX(t) \tag{7.4}$$

其中最优控制向量

$$K = R^{-1}B^TP \tag{7.5}$$

矩阵 P 必须满足下列方程：

$$A^TP + PA - PBR^{-1}B^TP + Q = 0 \tag{7.6}$$

式（7.5）称为退化矩阵黎卡提方程。设计步骤如下：

1）解退化矩阵黎卡提方程式（7.6）求出矩阵 P，如果存在正定矩阵 P（某些系统可能没有正定矩阵 P），那么系统是稳定的，即矩阵 $A-BK$ 是稳定矩阵。

2）将矩阵 P 代入式（7.5），求得的矩阵 K 就是最优矩阵。

MATLAB 有一条命令 lqr，它给出连续时间黎卡提方程的解，并能确定最佳反馈增益矩阵。

7.1.2 LQR Control Algorithm

Suppose the controlled object is an inverted pendulum system, with its system state equation

$$\begin{cases} \dot{X} = AX + Bu \\ Y = CX + Du \end{cases} \tag{7.7}$$

7.1.2 LQR 控制算法

假设被控对象是倒立摆系统，系统状态方程为

$$\begin{cases} \dot{X} = AX + Bu \\ Y = CX + Du \end{cases} \tag{7.7}$$

And suppose that the coefficient matrices of state equation are as follows:

$$A=\begin{pmatrix}0 & 1 & 0 & 0 & 0 & 0\\ -2272.7 & -0.0018 & 2272.7 & 0 & 0 & 0\\ 0 & 0 & 0 & 1 & 0 & 0\\ 2207.5 & 0 & -2207.5 & -0.1766 & 0.8437 & 0\\ 0 & 0 & 0 & 0 & 0 & 1\\ 6622.5 & 0 & -6622.5 & -0.5298 & 31.931 & 0\end{pmatrix}$$

$$B=\begin{pmatrix}0\\0.9091\\0\\0\\0\\0\end{pmatrix} C=\begin{pmatrix}1&0&0&0&0&0\\0&0&1&0&0&0\\0&0&0&0&1&0\end{pmatrix} D=\begin{pmatrix}0\\0\\0\end{pmatrix} \quad (7.8)$$

LQR is based on a controllable system.

According to system controllability criterion we could get rank(Mc) =6, the system is controllable. So, adding a LQR controller to the original system makes the closed - loop system stable and satisfy the transient performance indicators. The main purpose of the design of controller by using the LQR is to get feedback gain matrix K. When designing a state feedback controller, one of the most important problems is the selection of weighted matrices Q and R in the quadratic performance index function. In order to simplify the problems and make the weighted matrices have clearer physical meanings, we have a diagonal matrix as Q.

$$Q=\begin{pmatrix}Q_{11}&0&0&0&0&0\\0&Q_{22}&0&0&0&0\\0&0&Q_{33}&0&0&0\\0&0&0&Q_{44}&0&0\\0&0&0&0&Q_{55}&0\\0&0&0&0&0&Q_{66}\end{pmatrix}, R=(r) \quad (7.9)$$

状态方程系数矩阵如下

$$A=\begin{pmatrix}0 & 1 & 0 & 0 & 0 & 0\\ -2272.7 & -0.0018 & 2272.7 & 0 & 0 & 0\\ 0 & 0 & 0 & 1 & 0 & 0\\ 2207.5 & 0 & -2207.5 & -0.1766 & 0.8437 & 0\\ 0 & 0 & 0 & 0 & 0 & 1\\ 6622.5 & 0 & -6622.5 & -0.5298 & 31.931 & 0\end{pmatrix}$$

$$B=\begin{pmatrix}0\\0.9091\\0\\0\\0\\0\end{pmatrix} C=\begin{pmatrix}1&0&0&0&0&0\\0&0&1&0&0&0\\0&0&0&0&1&0\end{pmatrix} D=\begin{pmatrix}0\\0\\0\end{pmatrix} \quad (7.8)$$

LQR 控制的前提条件是系统是能控的。

由可控性判据可知 rank(Mc) =6，系统是可控的。因此可以尝试给系统加上 LQR 控制器使得系统闭环稳定，且满足暂态性能指标。在运用线性二次型最优控制算法进行控制器设计时，主要的目的就是获得反馈增益矩阵 K 的值。由上一小节的推导知道，设计系统状态反馈控制器时，一个关键的问题就是二次型性能指标泛函中加权矩阵 Q 和 R 的选取。为了使问题简化及加权矩阵具有比较明确的物理意义，我们将 Q 取为对角阵。

$$Q=\begin{pmatrix}Q_{11}&0&0&0&0&0\\0&Q_{22}&0&0&0&0\\0&0&Q_{33}&0&0&0\\0&0&0&Q_{44}&0&0\\0&0&0&0&Q_{55}&0\\0&0&0&0&0&Q_{66}\end{pmatrix}, R=(r) \quad (7.9)$$

The functional performance index is

$$J=\int_0^{\infty}\begin{bmatrix} Q_{11}x_1^{\ 2} + Q_{22}\dot{x}_1^2 + Q_{33}x_2^2 + \\ Q_{44}\dot{x}_2^2 + Q_{55}\theta^2 + Q_{66}\dot{\theta}^2 + ru^2 \end{bmatrix}\mathrm{d}t \tag{7.10}$$

From (7. 10) we can know Q_{ii} is the weight of square of x_i and the square of $\dot{x}_i$, Q_{ii} increases relatively means that the demand for x_i and $\dot{x}_i$ becomes strict relative to other state variables, the proportion of performance index is larger, the deviation of state x_i and $\dot{x}_i$ is less. r is the square weight of controlled quantity u, a relatively larger means control cost increases, making control energy smaller, and feedback is abate, when the value of r becomes small, the system control cost decreases, the feedback increases, dynamic responses of the system is quickly.

Considering the flexible connection level inverted pendulum system in the process of operation, the maincontrolled variables are system outputs x_1、x_2 and θ, so on selection of various components of the weighted diagonal matrix Q, choosing Q_{11} represents the weight of the driving car displacement, Q_{33} indicates the weight of the driven car displacement, and Q_{55} represents the weight of swinging rod angle. At the same time, in order to avoid the car speed and the swinging rod in larger angular velocity of the jitter, the value of Q_{22}、$Q_{44}$and Q_{66} also need to be set in a reasonable range.

The design of the control system is shown in Figure 7. 1, in which R is the step input

性能指标泛函为：

$$J=\int_0^{\infty}\begin{bmatrix} Q_{11}x_1^{\ 2} + Q_{22}\dot{x}_1^2 + Q_{33}x_2^2 + \\ Q_{44}\dot{x}_2^2 + Q_{55}\theta^2 + Q_{66}\dot{\theta}^2 + ru^2 \end{bmatrix}\mathrm{d}t \tag{7.10}$$

由（7. 10）式可以看出，Q_{ii}是对 x_i 及$\dot{x}_i$ 的平方的加权，Q_{ii}相对增加就意味着对 x_i 及$\dot{x}_i$ 的要求相对其他状态变量严格，在性能指标中的比重变大，x_i 及$\dot{x}_i$ 的偏差状态相对减小。r 是对控制量 u 的平方加权，当 r 相对较大时，意味着控制费用增加，使得控制能量较小，反馈减弱，而 r 取值较小时，系统控制费用减小，反馈增强，系统动态响应迅速。

考虑到一级柔性连接倒立摆系统在运行过程中，主要的被控量为系统的输出量 x_1、x_2 和 θ，因此在选取加权对角阵 Q 的各元素值时，选择 Q_{11}代表主动小车位移的权重，Q_{33}代表从动小车位移的权重，而 Q_{55}代表摆杆角度的权重；同时，为避免小车的速度和摆杆角速度出现较大的抖动，也要将 Q_{22}、Q_{44}和 Q_{66}的值设置在合理的范围内。

控制系统如图 7. 1 所示，图中 R 是施加在小车上的阶跃输入，六个状态量

applied on the car, six state quantity x_1, $\dot{x}_1$, x_2, $\dot{x}_2$, θ and $\dot{\theta}$ represent the driving car displacement, driving car speed, driven car displacement and driven car speed, angle and angular velocity of swinging rod, the output $y=[x_1, x_2, \theta]'$ includes driving car displacement and driven car displacement and angle of swinging rod. We want to design a controller that the swinging rod will swing with a step input on the system, then return to the vertical position, finally, the car arrived at the new command position.

x_1、$\dot{x}_1$、x_2、$\dot{x}_2$、θ 和 $\dot{\theta}$ 分别代表主动小车位移、主动小车速度、从动小车位移、从动小车速度、摆杆角度和摆杆角速度，输出 $y=[x_1, x_2, \theta]'$ 包括主动小车位移、从动小车位移和摆杆角度。我们要设计一个控制器，使得当给系统施加一个阶跃输入时，摆杆会摆动，然后仍然回到垂直位置，小车到达新的命令位置。

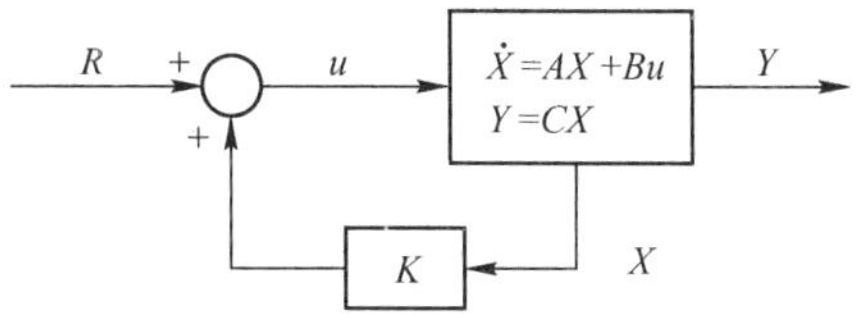

Figure 7.1 The LQR Control Principle Diagram

图 7.1 LQR 控制原理图

7.2 The Simulation of LQR Controller

7.2 LQR 控制器的仿真

7.2.1 Weight Q's Influence on The System Responses

From the MATLAB simulation analysis, we can get the system simulation curve with different Q. The simulation process is to get the system output curve when the step response of system is 0.1.

(1) When $Q_{11}=1$, $Q_{22}=1$, $Q_{33}=1$, $Q_{44}=1$, $Q_{55}=1$, $Q_{66}=1$, $R=1$

1) The feedback gain matrix K

From the MATLAB program Stage1_LQR.m, we could get the feedback gain matrix:

7.2.1 权值 Q 对系统响应的影响

对系统进行 MATLAB 仿真分析，可以得到当 Q 不同时的系统仿真曲线。仿真过程是对系统施加 0.1 的阶跃响应，得到系统的输出曲线。

(1) 当 $Q_{11}=1$，$Q_{22}=1$，$Q_{33}=1$，$Q_{44}=1$，$Q_{55}=1$，$Q_{66}=1$，$R=1$ 时

1) 反馈增益矩阵 K

利用 MATLAB 程序 Stage1_LQR.m，求得反馈增益矩阵：

$K = [103.4705 \quad 14.9220 \quad -104.8847 \quad -18.8044 \quad 57.6835 \quad 10.5030]$

2) The system step response simulation curve is shown in Figure 7.2.

$K = [103.4705 \quad 14.9220 \quad -104.8847 \quad -18.8044 \quad 57.6835 \quad 10.5030]$

2）系统阶跃响应仿真曲线如图 7.2 所示。

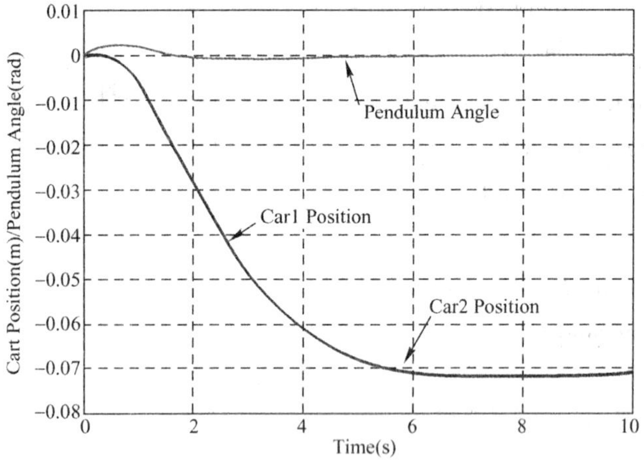

Figure 7.2 The Closed – Loop Step Response Simulation Curve of One – Stage Pendulum LQR Control System

图 7.2 一级摆 LQR 控制系统的闭环阶跃响应仿真曲线
（$Q_{11}=1$，$Q_{22}=1$，$Q_{33}=1$，$Q_{44}=1$，$Q_{55}=1$，$Q_{66}=1$）

As we can see from Figure 7.2, the displacement curve of driving car and driven car almost overlap, system response overshoot volume is small, but stable time and rise time is big, the cars moves in the opposite direction.

从图 7.2 中可以看出，主动小车和从动小车的位移曲线几乎重合，系统响应的超调量很小，但稳定时间和上升时间偏大，小车朝相反的方向移动。

(2) When $Q_{11}=1000$, $Q_{22}=1$, $Q_{33}=1000$, $Q_{44}=1$, $Q_{55}=100$, $Q_{66}=1$, $R=1$

1) Feedback gain matrix K

Run the program to get

$K = [194.7408 \quad 20.5237 \quad -239.4621 \quad -51.9351 \quad 128.9268 \quad 23.5417]$

2) The system step response simulation curve is shown as figure 7.3.

（2）当 $Q_{11}=1000$，$Q_{22}=1$，$Q_{33}=1000$，$Q_{44}=1$，$Q_{55}=100$，$Q_{66}=1$，$R=1$ 时

1）反馈增益矩阵 K

执行程序得到：

$K = [194.7408 \quad 20.5237 \quad -239.4621 \quad -51.9351 \quad 128.9268 \quad 23.5417]$

2）系统阶跃响应曲线如图 7.3 所示。

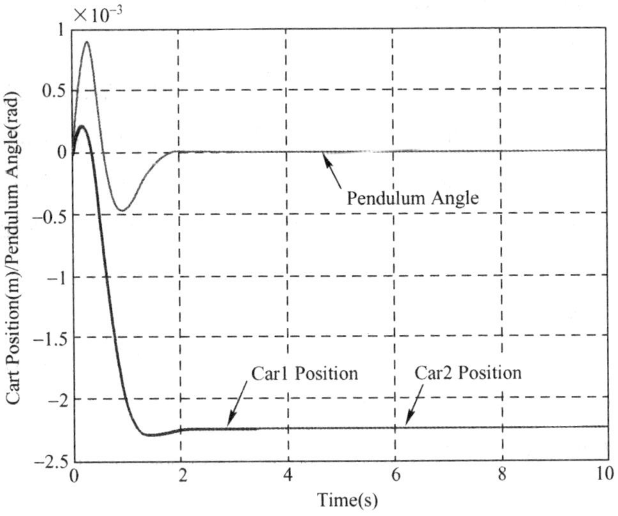

Figure 7.3 The Closed－Loop Step Response Simulation Curve of One－Stage Pendulum LQR Control System

图 7.3 一级摆 LQR 控制系统的闭环阶跃响应仿真曲线

($Q_{11}=1000$，$Q_{22}=1$，$Q_{33}=1000$，$Q_{44}=1$，$Q_{55}=100$，$Q_{66}=1$，$R=1$)

If adjust the Q_{11}、Q_{33} and Q_{55} in large range, the response of the system will also improve. But under the condition of Q_{11}, Q_{33} and Q_{55} are small enough, system response meets the requirements.

如果再增大 Q_{11}，Q_{33} 和 Q_{55}，系统的响应还会改善。但在保证 Q_{11}，Q_{33} 和 Q_{55} 足够小的情况下，系统响应已经满足要求了。

After repeatedly changing values of Q and R, we found that the following four aspects need to be pay attention to selecting its value:

通过反复调整权值 Q 和 R 发现，在选取这些值的时候需要注意以下四个方面：

1) Because system model is linearized, in order to make the system state variable work in the linear range, each state volume should not be too large.

1）由于采用的系统模型是线性化后的结果，为使系统各状态量能够在线性范围工作，要求各状态量不宜过大。

2) The closed－loop system should have a pair of conjugate complex poles, it helps to overcome the system's nonlinear friction, but the mode of the dominant pole should not be too big in case that system has

2）闭环系统最好能有一对共轭复数极点，这样有利于克服系统的非线性摩擦，但系统主导极点的模不应太大以免系统频带过宽，使得系统对噪声过于敏感，导致系统不能正常工作。

over wide frequency band, which makes the system so sensitive to noise and can't work normally.

3) The change of the weighted matrix R can lead to the system consumes a lotof control energy, we should control the size of $U(t)$ that can't exceed the actuator system ability, making the amplifier in the saturated state.

4) As the main controlled volume, the weight of system output x_1, x_2 and θ should be greater than the weight of $\dot{x}_1$, $\dot{x}_2$ and $\dot{\theta}$.

3）加权矩阵 R 的变化，会导致系统消耗大量的控制能量，应注意控制 $U(t)$ 的大小，不能超过系统执行机构的能力，使得放大器处于饱和状态。

4）由于系统的输出量中 x_1、x_2 和 θ 为主要的被控量，因此其权值应大于 $\dot{x}_1$、$\dot{x}_2$ 和$\dot{\theta}$的权值。

7.2.2 Eliminate the Steady - StateError

In the previous design method, the system has steady - state error, mainly because the output signal feedback multiplied by a feedback gain matrix K, and then directly subtracted from input and the control signal. This leads to the different dimensions of input and feedback, therefore, in order to make the input and feedback of dimensional match each other, the input should be multiplied by the gain $\overline{N}$, as shown in Figure 7.4.

7.2.2 消除稳态误差

在前面的设计方法中，系统存在稳态误差，主要是因为把输出信号反馈回来乘以一个反馈增益矩阵 K，然后直接与输入量相减得到控制信号。这就使得输入与反馈的量纲不一致，因此，为了使输入与反馈的量纲匹配，给输入乘以增益$\overline{N}$，如图 7.4 所示：

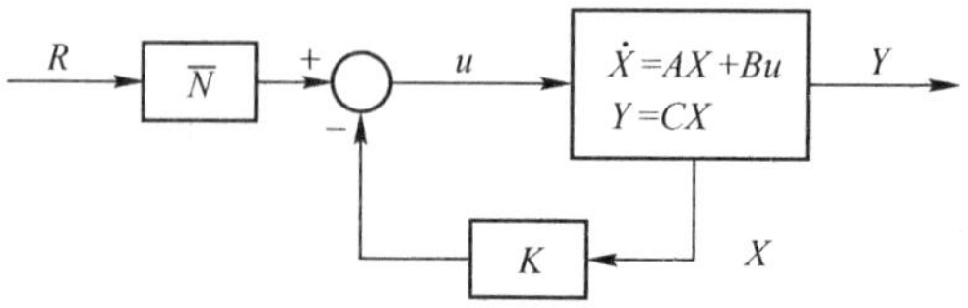

Figure 7.4 Dimensional Matching LQR Control Principle Diagram

图 7.4 量纲匹配的 LQR 控制原理图

The dimension matched LQR control MATLAB program is Stage 1_LQR_N bar. m. This article uses the input and output function rscale for matching coefficient, it is not a commonly used MATLAB function, so we need to put rscale. m and source program to the MATLAB workspace together.

具有量纲匹配的 LQR 控制 MATLAB 程序为 Stage1_LQR_Nbar. m，由于本文中要用到求取输入输出匹配系数函数 rscale，它不是 MATLAB 常用函数，因此必须把程序 rscale. m 和源程序一起复制到 MATLAB 工作区中。

Running the program Stage 1_LQR_N bar. m, we get the feedback gain matrix

$K=[194.7408 \quad 20.5237 \quad -239.4621 \quad -51.9351 \quad 128.9268 \quad 23.5417]$

$\overline{N}=\text{rscale}(A,B,Cn,0,K)=-194.7408$, It will be known, $\overline{N}$ is equal to the car position x_1 corresponding to the vector K.

运行程序 Stage1_LQR_Nbar. m，得到反馈增益矩阵

$K=[194.7408 \quad 20.5237 \quad -239.4621 \quad -51.9351 \quad 128.9268 \quad 23.5417]$

$\overline{N}=\text{rscale}(A,B,Cn,0,K)=-194.7408$，由此可知，$\overline{N}$ 与 K 向量中小车位置 x_1 对应的那一项相等。

The system response curve is shown in Figure 7.5, the driving car and driven car displacement can track the input signal, and the swinging rod overshoot is small enough, rise time and stability time is lesser as well.

此时系统的响应曲线如图 7.5 所示，主动小车位移和从动小车位移均能跟踪输入信号，并且摆杆超调足够小，上升时间和稳定时间也较小。

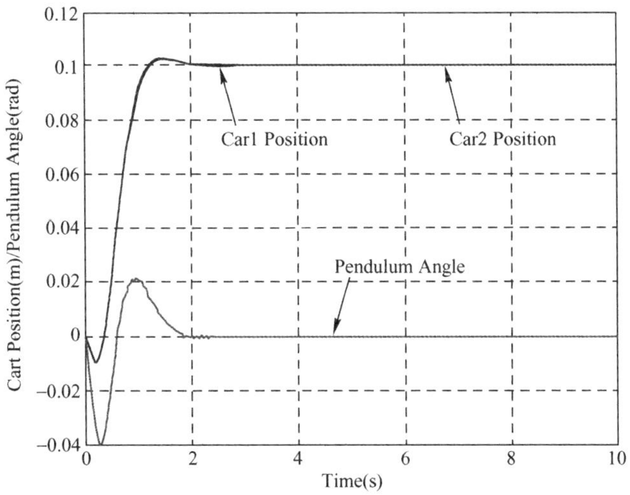

Figure 7.5 The Closed Loop Step Response Simulation Curve of One – Stage Pendulum Dimension Matching LQR Control System

图 7.5 一级摆量纲匹配的 LQR 控制系统的闭环阶跃响应仿真曲线

7.3 The Actual Control of LQR Controller

The flexible connection inverted pendulum system can be directly controlled by the MATLAB. Simulink can complete the real – time control of the controlled object.

The start module should be added in actual control operation. Driving car in accordance with the procedures the start module drives the driven car, and then drives the swinging rod swinging, the pendulum angle is stable in the process of movement. One – stage flexible connection inverted pendulum LQR actual system control block diagram is shown as Figure 7.6.

7.3 LQR 控制器的实际控制

柔性连接倒立摆系统可由 MATLAB 直接控制。通过 Simulink 即可完成对被控对象的实时控制。

可以在实际控制操作中加入起摆模块。主动小车按照起摆模块的程序，带动从动小车，进而带动摆杆运动，在运动的过程中使摆角达到稳定控制的范围内。一级柔性连接倒立摆 LQR 实际系统的控制框图如图 7.6 所示。

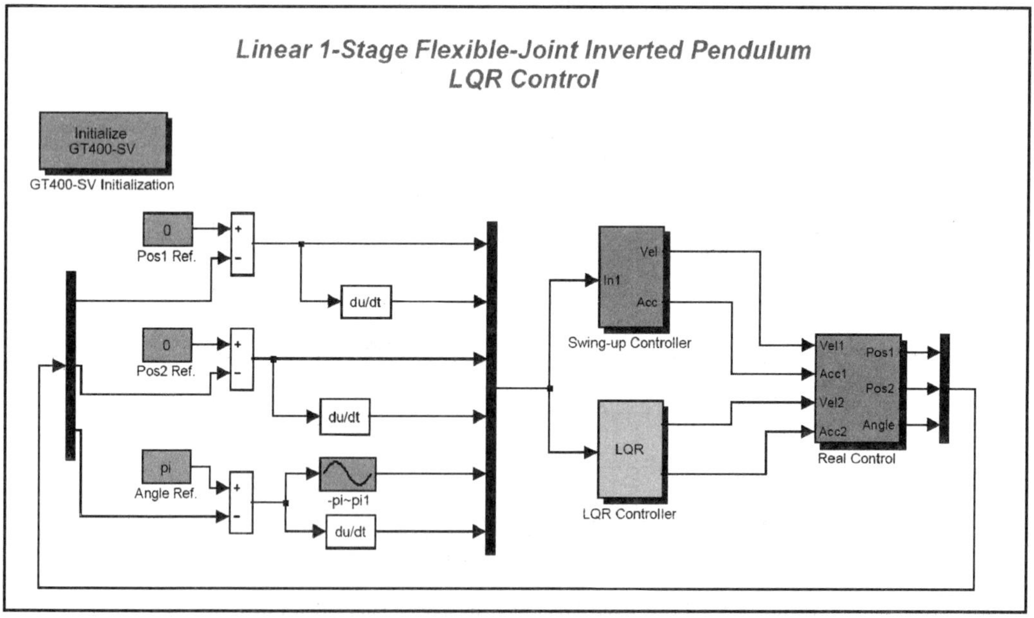

Figure 7.6 One – Stage Pendulum Actual Control Diagram

图 7.6 一级摆 LQR 实际系统的控制框图

Where "GT400 – SV Initialization" is a system initialization module, "Real Control" is real – time control module, which can realize the read and write function of the controlled variable and state variable. "Swing – up Controller" is the pendulum controller module. "LQR Controller" is LQR controller module, which can be set by Simulink module. Double – click to open the "LQR Controller", as shown in Figure 7.7.

其中，"GT400 – SV Initialization" 为系统初始化模块；"Real Control" 为实时控制模块，可实现系统控制量及状态量的读取和写入功能。"Swing – up Controller" 为起摆控制器模块，"LQR Controller" 为LQR控制器模块，由Simulink模块组建而成。双击打开 "LQR Controller"，如图7.7所示。

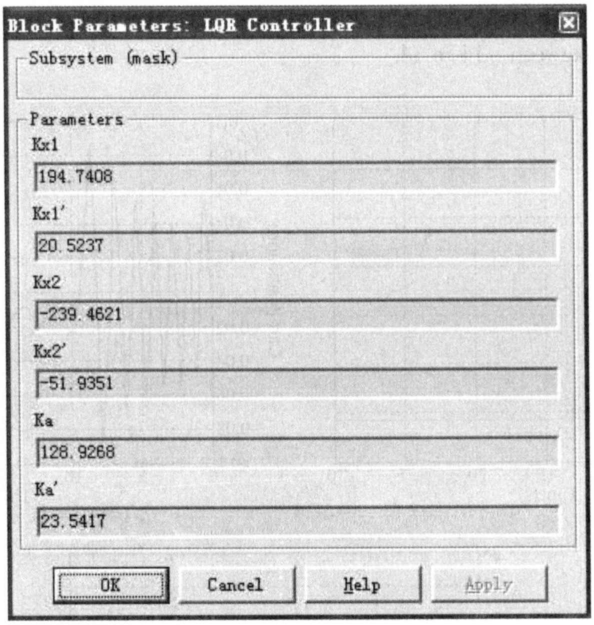

Figure 7.7 LQR Controller Parameter Settings Window

图7.7 LQR控制器参数设置窗口

When the feedback gain matrix is

$$K = [194.7408 \quad 20.5237 \quad -239.4621 \quad -51.9351 \quad 128.9268 \quad 23.5417],$$

we can experiment on one – stage flexible connection inverted pendulum system for disturbance state and steady state.

当反馈增益矩阵为：

$$K = [194.7408 \quad 20.5237 \quad -239.4621 \quad -51.9351 \quad 128.9268 \quad 23.5417],$$

可以对一级柔性连接倒立摆系统进行稳定状态实验和扰动状态实验。

As we can see from Figure 7.8, the LQR control strategy in one - stage flexible connection inverted pendulum system is effective. The inverted pendulum can keep balance in the predetermined position, and keep stable control. Overshoot of swinging rod angle is small, settling time is less than 2s, the steady state error is 0.3°.

从图 7.8 中可以看出，LQR 控制策略对一级柔性连接倒立摆系统的控制是有效的，倒立摆基本上能在指定位置保持平衡，并且控制效果平稳。其中摆杆角度超调量小，达到稳态时间不超过 2s，稳态误差为 0.3°。

Both the steady - state error of the two cars is 0.01 m, with small oscillation in equilibrium position. This is mainly due to measurement noise existing in the system and the nonlinear friction factors. The input of control is stable and always between -14 to 14.

两个小车的稳态误差大小均为 0.01 m，并在平衡位置做微小的振荡，这主要是因为系统中存在测量噪声以及非线性摩擦力等因素。控制量也比较平稳，始终保持在 -14 ~ 14 之间。

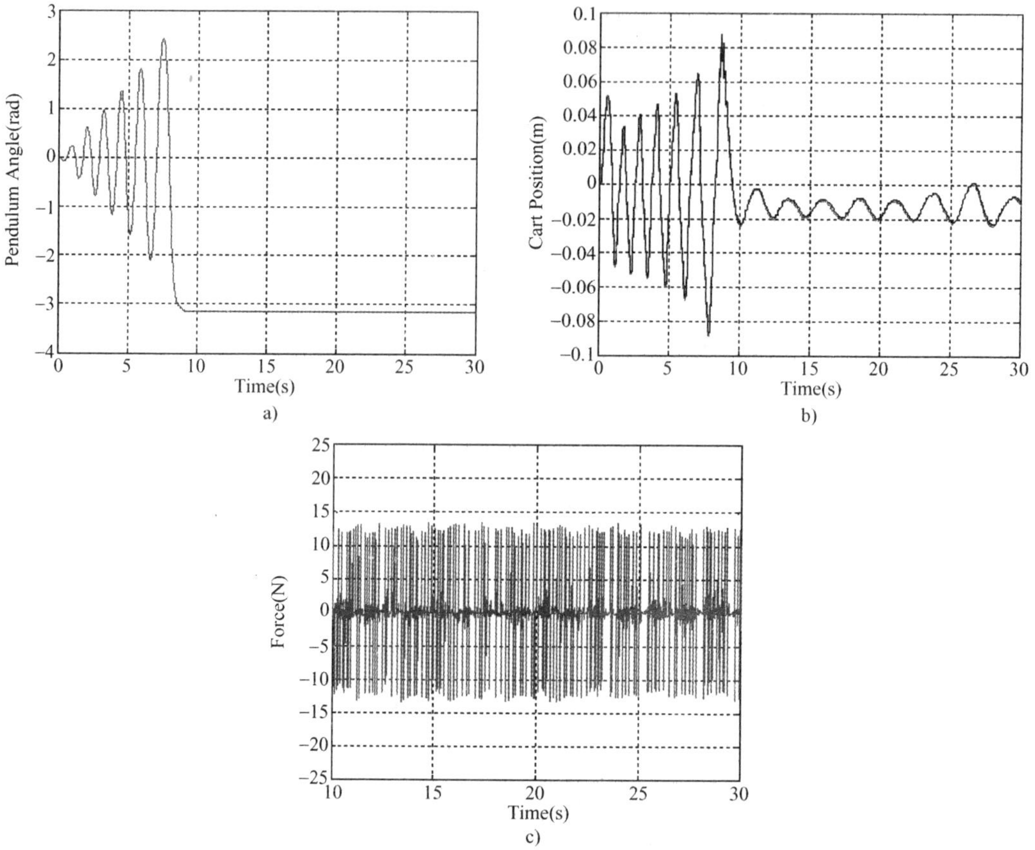

Figure 7.8 One - Stage Pendulum LQR Actual System Control Curve: Steady State Experiment

图 7.8 一级摆 LQR 实际系统的控制曲线：稳定状态实验

a) Swinging rod angle b) Driving car and driven car displacement c) Control law

a) 摆杆角度 b) 主动小车位移和从动小车的位移 c) 控制量

Repeat the experiment for many times, it has been found that because of the limitation of guide rail length of pendulum rod control system, the biggest controllable scope of the initial angle is 28°. For stable control, the biggest controllable scope of the angular is 16°.

通过多次实验发现，由于导轨长度的限制，控制系统对摆杆初始角度的可控最大范围是28°；稳定时，摆角的可控最大范围是16°。

State disturbe experiment means when the systems in stable, give a disturbance signal within a certain interval manually (touch the swinging rod with a pen quickly and lightly). As shown in Figure 7.9, by three times of disturbance respectively, swinging rod can quickly adjust to balance position, it shows that system has better robust performance. Figure 7.10 is a photo of flexible connection level inverted pendulum system in stable condition.

对系统进行扰动状态实验是指在系统稳定时人为加入扰动量（用笔杆轻碰摆杆顶部）。图7.9中显示摆杆在分别受到三次扰动后，能够迅速调整至平衡位置，说明系统具备较好的鲁棒性能。

图7.10为一级柔性连接倒立摆系统稳定时的照片。

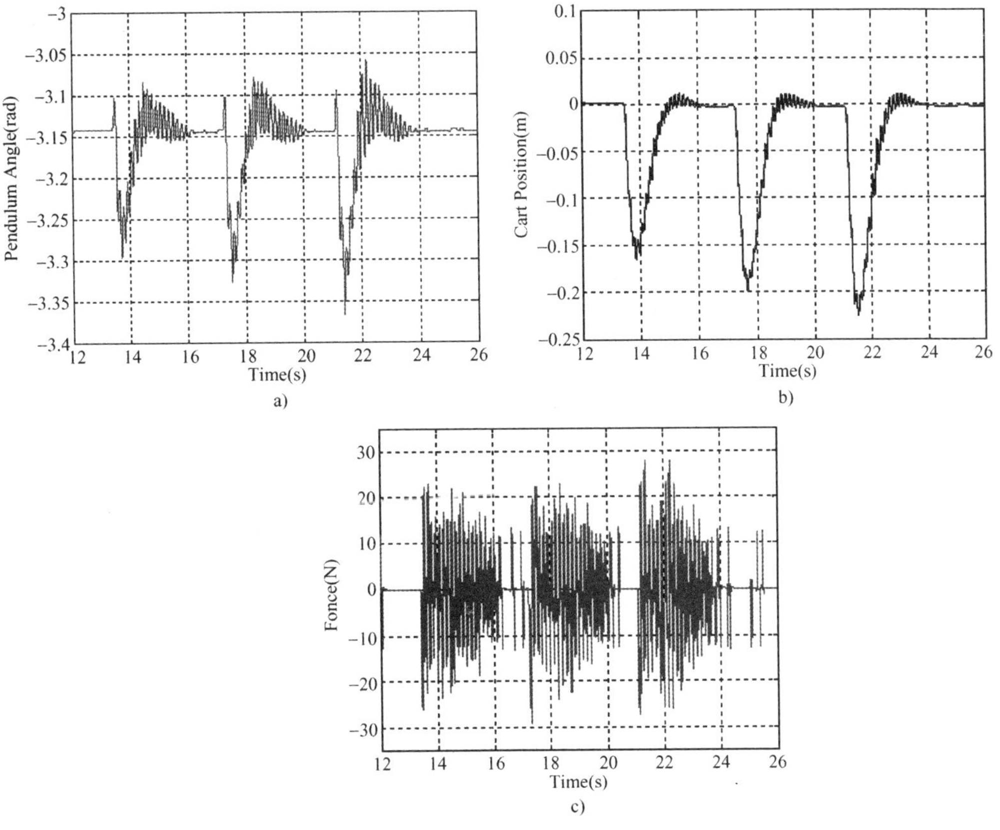

Figure 7.9 One – Stage LQR Actual System Control Curve: Disturbance State Experiment

图7.9 一级摆LQR实际系统的控制曲线：扰动状态实验

a) Swinging rod angle b) Driving car and driven car displacement c) Control Signal

a）摆杆角度 b）主动小车位移和从动小车的位移 c）控制信号

Figure 7. 10　One - Stage Pendulum System in Steady State
图 7. 10　一级摆系统稳定时的照片

7. 4　The Experimental Steps

1) Push the driving car and the driven car to the middle position of the rails, and make the swinging rod in the stationary state of free prolapsed.

2) Open the MATLAB software in the computer, start the electric cabinet power supply.

3) Open the one - stage pendulum LQR actual system control block diagram.

4) Click the " compile" button, and click "connect" button after successful compilation, then click "run" button, the flexible connection inverted pendulum system will automatically move, and switch to the LQR control at the proper angle, ultimately remain stable.

5) If the control effect is not ideal, adjust parameters until it achieves satisfied condition.

7. 4　实验步骤

1）将主动小车和从动小车推到导轨正中间位置，并且使摆杆处于自由下垂的静止状态。

2）打开 MATLAB 软件，再打开电控箱电源。

3）打开一级摆 LQR 实际系统的控制框图。

4）点击“编译”按钮，编译成功后点击“连接”按钮，再点击“运行”按钮，柔性连接倒立摆系统将自动起摆，并在适当角度切换至 LQR 控制，最终保持稳定。

5）如果控制效果不理想，可适当调整参数，直到获得较好的控制效果。

7.5 Experiment Instruction

7.5 实验指导书

姓名（Name）:　　　　　　　　院系（School）:
专业（Major）:　　　　　　　　学号（Student ID）:
起至日期:　年　月　日　至　年　月　日
from:　Y　M　D　to　Y　M　D

实验题目：一级柔性连接倒立摆系统的控制器设计

Experimental Title: Controller Design for One - Stage Flexible Connection Inverted Pendulum

Ⅰ. The technical parameters

Controlled object: one - stage flexible connected inverted pendulum GLIP2011 of Googol Company.

Relevant parameters:

M_1 driving car mass 1.1 kg;

M_2 driven car mass 1.1 kg;

ms winging rod mass 0.13 kg;

l length between center of rotationand mass of pendulum 0.25 m;

k the coefficient of stiffness of spring 2500 N/m;

c_1 friction coefficient between driving car and guide rail 0.2 N · s/m;

c_2 friction coefficient between driven car and guide rail 0.2 N · s/m;

T sampling time 0.005 s.

Ⅱ. Experimental requirements

1. Deduce the state space equation and analyze the stability and controllability;
2. Design the optimal controller with the following indexes:
   1) The settling time is less than 3 second;
   2) The amplitude of θ is less than 0.3°;
   3) The steady - state error is less than 2%.

Ⅲ. Experimental content

1. Set up the linear mathematical model of flexible connected inverted pendulum;
2. Design, simulation and actual control of the optimal controller for one - stage flexible connected inverted pendulum system;
3. Print the output of experiment;
4. Analyze the experiment results.

一、技术参数：

本系统设计的被控对象采用固高公司的直线一级柔性连接倒立摆系统 GLIP2011。

系统内部各相关参数为：

M_1 主动小车的质量 1.1 kg;

M_2 从动小车的质量 1.1 kg;

m 摆杆质量 0.13 kg;

l 摆杆转动中心到杆质心的长度 0.25 m;

k 弹簧的倔强系数 2500 N/m;

c_1 主动小车与导轨之间的摩擦系数 0.2 N · s/m;

c_2 从动小车与导轨之间的摩擦系数 0.2 N · s/m;

T 采样时间 0.005 s。

二、设计要求：

1. 推导出系统的状态空间方程。用 MATLAB 进行稳定性及可控性分析；
2. 设计最优控制器，使得闭环系统的响应指标为：
   1）系统过渡过程时间小于 3 s;
   2）θ 的幅值小于 0.3°;
   3）稳态误差小于 2%。

三、实验内容

1. 建立柔性倒立摆的线性数学模型；
2. 对一级柔性倒立摆系统的最有控制器进行设计、仿真与实物实验；
3. 打印实验结果；
4. 分析实验结果。

7.6 Requirements for the Experiment Report

1) Show the result of MATLAB program execution and the simulation curve;

2) Get the actual controller parameters and control curve, and state whether the performance index meet the requirements or not.

3) Give the response curve of steady state experiment and the disturbed state experiment.

7.6 实验报告要求

1）给出 MATLAB 程序的执行结果和仿真曲线；

2）给出实际控制器参数和控制曲线，并说明性能指标是否满足要求；

3）给出系统对稳定状态实验和扰动状态实验的响应曲线。

Chapter 8 Fuzzy Controller Design for Ball-beam System

第8章 针对球杆系统的模糊控制器设计实验

In this chapter, the mathematical modelling by Lagrange equation for ball - beam system will be introduced first. Although the Ball - beam system is a nonlinear system, in the small angle domain, the location control will be considered as a linear control. By means of Taylor series method for linearization, the linear mathematical model will be set up. This linear model is not only convenient for simulation and adjustment of parameters, but also easy for ball - beam system controller design.

本章首先介绍通过拉格朗日方程建模方法建立球杆系统的数学模型。因为球杆系统的数学模型具有非线性，而球杆系统在小角度范围内的位置控制可以看作是线性控制。运用泰勒级数法对其模型进行线性化，建立球杆系统简洁而直观的数学模型，从而为控制器的仿真及参数调整打下一个很好的基础，进而便于实际控制球杆系统。

Fuzzy control theory is developed on the basis of Fuzzy sets, it is a new controller method based on fuzzy sets, fuzzy language variable, fuzzy logic reasoning. Through the fuzzy logic and fuzzy reasoning, the human beings experiment will be described by mathematical model, the control rules will be deal with by computer automatically. The hard job will be replaced by computer. Fuzzy control is a kind of method which is suitable for industry process control and huge system control. Especially for nonlinear control which include uncertain facts in these industry systems, the fuzzy control method will obtain better control performance.

模糊控制理论是在模糊集合理论的数学基础上发展起来的，是一种以模糊集合论、模糊语言变量、模糊逻辑推理为数学基础的新型计算机控制方法。它通过模糊逻辑和模糊推理，把人们的实践经验数学化，变成计算机可以理解的控制规则，从而代替人们进行有效的实时控制。模糊控制是一种适合工业生产过程和大系统的控制方法。特别对于无法建立模型和含有不确定性因素的非线性系统，模糊控制也能取得较好的控制效果。

As for the Ball - beam system have a nonlinear fact, it is obviously good solution for adopting fuzzy control law. In this chapter, the fuzzy control algorithm and modelling will be discussed in detail.

球杆系统本身含有非线性因素，采用模糊控制是一种比较好的控制策略，可以实现较好的动态性能和控制精度。本章将对模糊控制算法进行详细分析和设计。

8.1 Introduction to Ball - beam System

8.1 球杆系统简介

Ball - beam system is a popular equipment for automatic control subject, it is a nonlinear and unsteady system. There are a lot of nonlinear facts in the structure of ball - beam system, such as:

- Dead zone
- Nonlinear of DC motor and drive belt
- Discontinues of location measurement
- The surface of lead rail is not smooth, there is nonlinear resistance force.

The ball - beam system is called nonlinear unsteady system, because the ball location is unknow, although the angle of lead rail is fixed. For the lead rail with fixed angle, the ball will move to the bottom of the beam in some acceleration. Modelling and controller design for a nonlinear system is a challenge, it is a good research question in laboratory also. Ball - beam system is a good tool for researchers, for this kind of system has the characteristic of "nonlinear", "unsteady" but with simple structure and safe to use in experiment.

球杆系统是为自动控制原理等基础控制课程教学实验设计的实验设备，是一个非线性不稳定系统。其执行机构包括很多非线性特性，包括：

- 死区
- 直流电动机和带轮的传动非线性
- 位置测量的不连续性
- 导轨表面不是严格的光滑斜面，产生非线性阻力

称球杆系统是不稳定系统是因为即便导轨的仰角是固定的，小球的位置仍然是未知的。对于一个固定的导轨仰角，小球以一个固定的加速度运动直到停在导轨的底端。在控制领域中，非线性不稳定系统的建模和控制器的设计有许多需要克服的难点，有必要首先在实验室中研究。而球杆系统正是解决这种矛盾的最好的工具。它简单、安全并且具备了非线性不稳定系统所具有的重要的动态特性。

8.1.1 Structure of Ball-beam System

8.1.1 球杆系统组成

There are two parts in Ball - beam system. One is mechanism, another is IPM 100 smart servo system.

球杆系统主要由两大部分构成，一为球杆系统机械部分，二为 IPM100 智能伺服驱动系统。

Mechanism includes base, ball, beam, drive belt wheel, frame, and DC motor. It is shown in figure 8. 1. Its principle is as following: The actuator of ball – beam system is a DC motor, the motor drives the gear to rotate by a belt, which is driven by the connecting rod to rotate the bar, the controller controls rod angle by control the DC motor negative and positive rolling.

机械部分主要包括底座、小球、横杆、减速皮带轮、支架、电动机等。具体结构如图 8. 1 所示。其工作原理如下：球杆系统的执行机构为一电动机，电动机通过皮带带动齿轮转动，进而通过连杆带动横杆转动，因此控制器可以通过控制直流电动机的正反转来控制横杆的角度。

The ball is put in the lead rail of beam, there is a steel bar in one side and a displacement sensor in another side.

小球放置在横杆上，横杆的一侧为钢杆，另一侧为直线位移传感器。

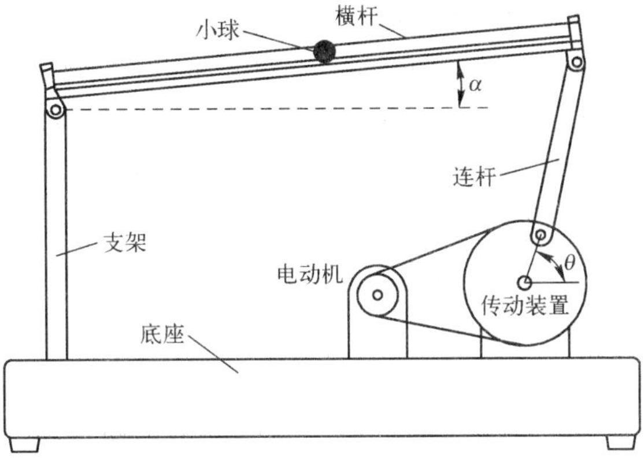

Figure 8. 1 Mechanism Structure of Ball – beam System

图 8. 1 球杆系统机械部分结构

When the ball runs in the lead rail, the position of ball in the rail will be measured by sensor in measurement bar, the output unit of sensor is voltage, which indicates the position variable x. When the real position of ball is not reach the input position, the beam will be driven by motor and keep a suitable angle with horizon. Meanwhile the ball will roll freely according to its weight vector, the ball position will be fixed by control the angle of lead rail. When the ball reaches the reference position, the beam rail will keep the angle α equal zero, the ball keep stop and system goes steady. The picture of Ball – beam system is shown in figure 8. 2.

当钢球在轨道上滚动时，通过测量钢杆上输出的电压信号获得钢球在轨道上的位置 x。当其实际位置与设定位置不符时，通过电动机转动使横杆与水平方向有一偏角。此时，小球重力分量可使其在导轨上自由滚动，通过对导轨角度的控制来使小球稳定在需要的位置。当小球达到设定位置时，此时保持导轨的偏角 α 为零度，小球停止运动，系统达到稳态。其实物照片如图 8. 2 所示。

Figure 8. 2 The Photo of Ball – beam System

图 8. 2 球杆系统实物图

The IPM100 smart servo driver is used to adjust DC motor. The IPM100 is a digital and intelligent controller, its driver has 100W circuit, fit for brush and brushless motor. According to the feedback control theory, when the sensor's signal is detected, after the signal processing, input a constant PWM voltage signal to wending of motor, a torque will act on the motor shaft, it makes the motor run. The torque should be set by the control algorithm in customer's program. There is a PID controller in IPM100 smart servo driver, the rotary position of motor will be controlled by the PID controller. Even the reference parameters of PID have been adjusted, it will guarantee that the motor has fast response and no percent overshoot.

IPM100 智能伺服驱动器的作用是控制电动机。它本身一个智能的高精度、全数字的控制器，内嵌 100W 的驱动电路，适合有刷和无刷电动机。基于反馈控制原理，在得到传感器信号后，对信号进行处理，然后给电动机绕组施加适当的 PWM 电压信号，一个相应的扭矩作用于电动机轴，使电动机开始运动。扭矩的大小决定于用户程序中的控制算法。IPM100 智能伺服驱动器内部包含一个 PID 控制算法，用来控制电动机转动的位置。PID 控制器的参数已经调整，保证电动机具有较快的响应并且没有超调。

8. 1. 2 The Structure of Ball – beam System

8. 1. 2 球杆系统控制结构

The principle and mechanism structure of Ball – beam system will be shown in Figure 8. 3.

基于球杆系统机械组成及其工作原理，其闭环控制系统结构如图 8. 3 所示。

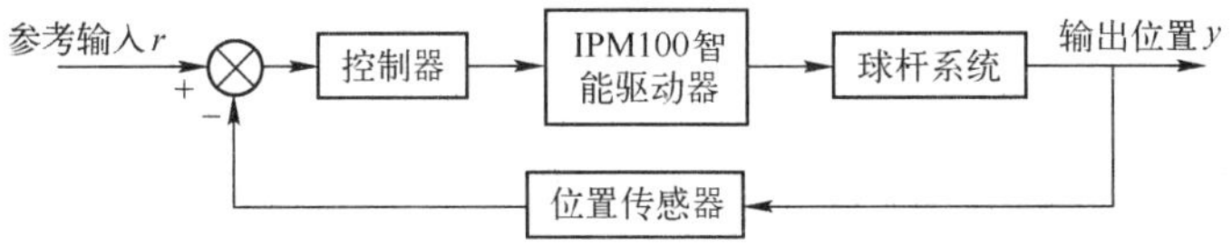

Figure 8. 3 The Control Diagram of Ball - beam System

图 8. 3 球杆系统控制结构图

It is obviously that the feedback control system is a constant control system according to the Figure 8. 3. the location of ball will be detected by position sensor, the error of sensor measurement data and control target value will be the input of controller. The output of controller is target angle of motor. And then the motor position will be adjusted by IPM100, so that the ball position will be adjusted also, until the ball arrive steady sate.

由系统控制结构图 8. 3 可以看出，该闭环系统为一典型恒值控制系统。通过位置传感器获取小球的实际位置，当与控制目标位置不符时，二者差值作为控制器的输入，控制器输出为电动机控制角度。之后由 IPM100 智能伺服驱动器根据控制角度调整电动机转动的位置，进而调整小球位置，直至达到稳态。

The flow chart of controller program will be shown in Figure 8. 4. and the control algorithm will be designed by user.

控制程序流程如图 8. 4 所示。其中控制算法可由用户自行设计。

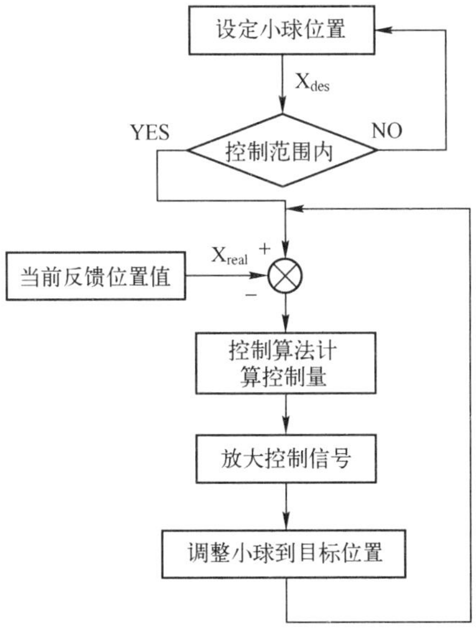

Figure 8. 4 The Program Chat of Ball - beam's Controller

图 8. 4 球杆系统控制程序流程图

8.2 The Modelling of Ball - beam System

8.2.1 The Analysis of Modelling Methods

For the mechanical system, we often use two kind of modelling methods: one is by means of Newton's law, another is by means of Lagrange equation. Newton's law is rely on the balance of force, it is very suitable for some simple structure mechanical systems which have no mutual constraints . But for the complex system, such as ball - bean system, which involves the transformation among multiple coordinate systems, and rquires a number of dynamic equations, it is more complex to use Newton's law. In contrast, the Lagrange equation is more suitable for the analysis of the motion of objects under the mutual constraints. Lagrange expression is very simple, It only needs to calculate the kinetic energy of the system and General force. Modelling of ball - beam system by Lagrange method avoids the complex force analysis in objects.

8.2.2 The Parameters of Ball - beam System

The parameters of ball - beam system is shown in Table 8. 1

8.2 球杆系统的数学模型建立

8.2.1 建模方法的选择

对于机械系统模型的建立，常用建模方法有两种：一种为牛顿定律建模方法，另一种拉格朗日方程建模方法。牛顿定律建立在力学平衡基础之上，对于简单的没有相互约束的机械系统，牛顿定律具有较好的建模效果。但对于一些复杂的系统，如球杆系统，应用牛顿定律进行分析，往往要涉及多个坐标系之间的变换，需要建立多个动力学方程，分析起来比较复杂。相比而言，拉格朗日方程更适合分析相互约束下物体的运动，其表达式非常简单，应用时只需计算系统的动能和广义力。用拉格朗日方法对球杆系统建模，避免了复杂的物体间的力学关系分析。

8.2.2 球杆系统的相关参数

球杆系统的参数见表 8. 1。

Table 8.1 The table of Ball – beam parameters

表 8.1 球杆系统各参数说明表

小球转动惯量 ball's inertia moment	J_b	球杆质量 beam's mass	M
小球半径 ball's radius	R	齿轮半径 gear's radius	d
小球质量 ball's mass	m	重力加速度 gravitional acceleration	g
小球在球杆上的位移 ball displacement on the beam	r	球杆转角 rotation angle of beam	α
球杆转动惯量 beam's inertia moment	J_l	齿轮转角 rotation angle of gear	θ
球杆长度 length of beam	L	作用在球杆上的扭矩 torque acting on the beam	τ

8.2.3 Modelling of Ball – beam System by Lagrange Method

The principle of Lagrange modelling method is described as following:

For a mechanical system, it will be described by the general coordinate system as

$$q^T=[q_1,q_2,\dots\dots,q_n]$$

$$\dot{q}^T=[\dot{q}_1,\dot{q}_2,\dots\dots,\dot{q}_n]$$

Assume that the total kinetic energy of the system is $V(q,\dot{q})$, total potential energy is $U(q)$, so the motion characteristics of the system can be described by Lagrange equation:

$$\frac{\mathrm{d}}{\mathrm{d}t}\left(\frac{\partial L}{\partial \dot{q}_i}\right)-\frac{\partial L}{\partial q_i}=\tau_i, i=1,2,\dots,n \tag{8.1}$$

Where τ_i is the sum of external forces or external moments in the i generalized coordinates q_i direction. L is the Lagrange function defined by following function:

$$L(q,\dot{q})=V(q,\dot{q})-U(q) \tag{8.2}$$

8.2.3 拉格朗日方法建立球杆系统数学模型

拉格朗日方程建模方法原理概括为：

对一个机械系统可以用广义坐标来描述：

$$q^{\mathrm{T}}=[q_1,q_2,\dots\dots,q_n]$$

$$\dot{q}^{\mathrm{T}}=[\dot{q}_1,\dot{q}_2,\dots\dots,\dot{q}_n]$$

记系统的总动能为 $V(q,\dot{q})$，总体势能为 $U(q)$。系统的运动特性可以用拉格朗日方程来描述：

$$\frac{\mathrm{d}}{\mathrm{d}t}\left(\frac{\partial L}{\partial \dot{q}_i}\right)-\frac{\partial L}{\partial q_i}=\tau_i, i=1,2,\dots,n \tag{8.1}$$

其中 τ_i 为第 i 个广义坐标 q_i 方向上的外部力或外部力矩之和。L 为下式定义的拉格朗日函数：

$$L(q,\dot{q})=V(q,\dot{q})-U(q) \tag{8.2}$$

When we set up the mathematical model of Ball – beam system, there are consisted of three part: mechanism model, angle model, and DC motor model. The relation between ball position in beam $r(t)$ and the beam angle $\alpha(t)$ will be combined by mechanism part. The angle model will combine the beam angle $\alpha(t)$ with the gear rotate angle $\theta(t)$. The Ball – beam system is controlled by IPM100 smart servo driver, the motor model will combine the input voltage $u(t)$ and the gear rotate angle $\theta(t)$. There is a PID controller in IPM100, the motor position will be controlled precisely, so we don't analysis the DC motor model here.

建立球杆系统模型，可将其分为三部分：球杆机械部分模型、角度模型和电动机模型。球杆机械部分模型将小球在横杆上的位置 $r(t)$ 和横杆的仰角 $\alpha(t)$ 联系起来，而角度模型则将横杆的仰角 $\alpha(t)$ 与齿轮的转动角度 $\theta(t)$ 联系起来。球杆系统由 IPM100 智能伺服驱动器进行控制，其电动机模型将电动机控制系统的输入电压 $u(t)$ 与齿轮的转动角度 $\theta(t)$ 联系起来。由于 IPM100 智能伺服驱动器内部含有 PID 控制器，已经精确控制电动机位置，对电动机模型没有必要详细分析，故不再叙述。

The mechanism part will be modeled by Lagrange equation. r, α, are the displacement from ball position to fixed end of the beam and the guide rail rotate angle along counter clock wise from horizon to new position. $\dot{r}$ and $\dot{\alpha}$ are placement velocity and angular velocity respectively. It is shown in figure 8.5.

对于球杆系统机械部分可以用拉格朗日方法建模。定义 r,α，分别表示小球相对固定端的位移和导轨相对水平位置沿固定轴逆时针旋转的角度。$\dot{r}$ 和 $\dot{\alpha}$ 分别表示相应的位移速度和角速度。如图 8.5 所示。

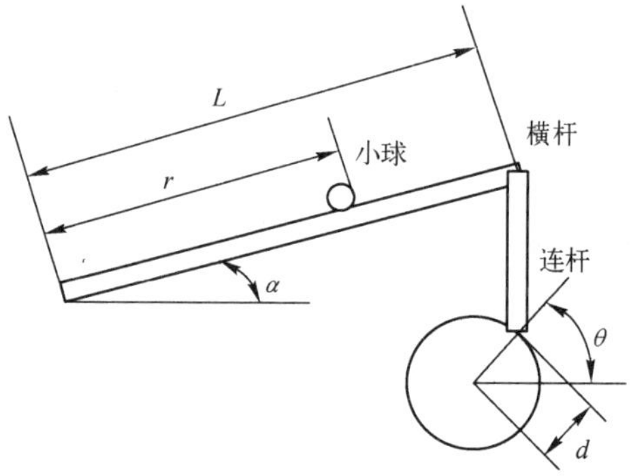

Figure 8.5 The Model of Ball – beam Diagram

图 8.5 球杆系统模型示意图

We analyze the kinetic energy firstly. The position of ball in rectangular coordinate system is (x,y), x, y have the relation as following with the general coordinate: $x = r\cos\alpha, y = r\sin\alpha$。Derivation of position function

$$\dot{x} = \dot{r}\cos\alpha - r\dot{\alpha}\sin\alpha \quad (8.3)$$

$$\dot{y} = \dot{r}\sin\alpha + r\dot{\alpha}\cos\alpha \quad (8.4)$$

The kinetic energy of ball along the r direction will be:

$$T_1 = \frac{1}{2}m(\dot{x}^2 + \dot{y}^2) = \frac{1}{2}m(\dot{r}^2 + r^2\dot{\alpha}^2) \quad (8.5)$$

When the ball rolls along the guide rail, there is self – rotation also. The relation between position and angle is as following:

Assume $\lambda = \frac{r}{R}$, λ is the self – rotation angle of the ball, R is the radius of the ball, r is the ball's displacement on the beam. So the angular velocity of ball is $\dot{\lambda} = \frac{\dot{r}}{R}$. The rotational kinetic energy of the ball is as following:

$$T_2 = \frac{1}{2}J_b\dot{\lambda}^2 = \frac{1}{2}J_b\frac{\dot{r}^2}{R^2} \quad (8.6)$$

Where J_b is the moment inertia

$J_b = \frac{2}{5}mR^2$.

Force analysis in the generalized coordinates:

The beam rotates around the fixed point, there is a kinetic energy:

$$T_3 = \frac{1}{2}J_l\dot{\alpha}^2 \quad (8.7)$$

Where J_l is moment inertia, $J_l = \frac{1}{3}ML^2$, so the sum of kinetic energy is:

$$T = T_1 + T_2 + T_3 \quad (8.8)$$

分析系统动能，小球在直角坐标系下的位置为(x,y)，x，y与广义坐标的关系为：$x = r\cos\alpha, y = r\sin\alpha$。对位置函数进行求导可得：

$$\dot{x} = \dot{r}\cos\alpha - r\dot{\alpha}\sin\alpha \quad (8.3)$$

$$\dot{y} = \dot{r}\sin\alpha + r\dot{\alpha}\cos\alpha \quad (8.4)$$

由此小球沿广义坐标r方向动能为：

$$T_1 = \frac{1}{2}m(\dot{x}^2 + \dot{y}^2) = \frac{1}{2}m(\dot{r}^2 + r^2\dot{\alpha}^2) \quad (8.5)$$

同时小球沿导轨方向滚动时，还有自身径向的转动。其转过角度与小球位置有如下关系：

设$\lambda = \frac{r}{R}$，λ为小球绕自身旋转的角度，R为小球的半径，r为小球在横杆上的位移。故小球角速度$\dot{\lambda} = \frac{\dot{r}}{R}$，因此小球绕径向转动动能为：

$$T_2 = \frac{1}{2}J_b\dot{\lambda}^2 = \frac{1}{2}J_b\frac{\dot{r}^2}{R^2} \quad (8.6)$$

其中J_b为小球的转动惯量。

$J_b = \frac{2}{5}mR^2$.

分析广义坐标下各坐标的受力：

球杆绕固定点旋转的动能大小为：

$$T_3 = \frac{1}{2}J_l\dot{\alpha}^2 \quad (8.7)$$

其中J_l为球杆的转动惯量，$J_l = \frac{1}{3}ML^2$，故系统的动能之和为：

$$T = T_1 + T_2 + T_3 \quad (8.8)$$

System potential energy analysis:

The ball's potential energy in the direction of r:

$$V_1 = mgr\sin\alpha \quad (8.9)$$

The mass of beam is M, so the potential energy of beam is as formula (8.10):

$$V_2 = \frac{1}{2}MgL\sin\alpha \quad (8.10)$$

Analysis of forces in general coordination:

There is no external force to the ball, and the ball's displacement is r, but the is beam is affected by the external torque τ.

So the Lagrange function will be:

$$L = T - V = T_1 + T_2 + T_3 - V_1 - V_2 \quad (8.11)$$

Rewrite the (8.11), there is:

$$L = \frac{1}{2}m(\dot{r}^2 + r^2\dot{\alpha}^2) + \frac{1}{2}J_b\dot{r}^2R + \frac{1}{2}J_l\dot{\alpha}^2 + mgr\sin\alpha - \frac{1}{2}Mgl\sin\alpha \quad (8.12)$$

In General coordinate there is $(q_1, q_2) = (r, \alpha)$,

$$\frac{\mathrm{d}}{\mathrm{d}t}\left(\frac{\partial L}{\partial \dot{q}_j}\right) = \begin{pmatrix} \left(\frac{J_b}{R^2} + m\right)\ddot{r} \\ (J_1 + mr^2)\ddot{\alpha} + 2m\dot{r}r\dot{\alpha} \end{pmatrix} \quad (8.13)$$

$$\frac{\partial L}{\partial q_j} = \begin{pmatrix} m\dot{r}\alpha^2 - mg\sin\alpha \\ -(mgr + \frac{1}{2}MgL)\cos\alpha \end{pmatrix} \quad (8.14)$$

$$\tau_j = \begin{pmatrix} 0 \\ \tau \end{pmatrix} \quad (8.15)$$

分析系统势能:

小球沿 r 方向引起的势能为:

$$V_1 = mgr\sin\alpha \quad (8.9)$$

球杆的质量为 M,则球杆的势能为式(8.10)所示:

$$V_2 = \frac{1}{2}MgL\sin\alpha \quad (8.10)$$

广义坐标下的受力分析:

小球位移 r,没有外力作用,而球杆转角 α 只受到外力施加的扭矩 τ。

综上,根据式(8.1)可得拉格朗日函数:

$$L = T - V = T_1 + T_2 + T_3 - V_1 - V_2 \quad (8.11)$$

改写式(8.11)获得下列方程:

$$L = \frac{1}{2}m(\dot{r}^2 + r^2\dot{\alpha}^2) + \frac{1}{2}J_b\dot{r}^2R + \frac{1}{2}J_l\dot{\alpha}^2 + mgr\sin\alpha - \frac{1}{2}Mgl\sin\alpha \quad (8.12)$$

广义坐标为:$(q_1, q_2) = (r, \alpha)$。故:

$$\frac{\mathrm{d}}{\mathrm{d}t}\left(\frac{\partial L}{\partial \dot{q}_j}\right) = \begin{pmatrix} \left(\frac{J_b}{R^2} + m\right)\ddot{r} \\ (J_1 + mr^2)\ddot{\alpha} + 2m\dot{r}r\dot{\alpha} \end{pmatrix} \quad (8.13)$$

$$\frac{\partial L}{\partial q_j} = \begin{pmatrix} m\dot{r}\alpha^2 - mg\sin\alpha \\ -(mgr + \frac{1}{2}MgL)\cos\alpha \end{pmatrix} \quad (8.14)$$

$$\tau_j = \begin{pmatrix} 0 \\ \tau \end{pmatrix} \quad (8.15)$$

As $\frac{\mathrm{d}}{\mathrm{d}t}\left(\frac{\partial L}{\partial \dot{q}_i}\right) - \frac{\partial L}{\partial q_i} = \tau_i, i = 1, 2, \ldots, n$, we will get:

$$\begin{cases} \left(\frac{J_b}{R^2} + m\right)\ddot{r} - mr\dot{\alpha}^2 + mg\sin\alpha = 0 \\ (J_l + mr^2)\ddot{\alpha} + 2mr\dot{r}\dot{\alpha} + \left(mgr + \frac{1}{2}MgL\right)\cos\alpha = \tau \end{cases} \tag{8.16}$$

We need the math relation between ball position and beam angle, so we only need the first formula in (8.16). And the expected angle of α trends to zero, so we will linearize α in the small size near zero. Then the formula (8.17) is obtained:

$$\ddot{r} = \frac{mg\alpha}{\frac{J_b}{R^2} + m} \tag{8.17}$$

After Laplace Transformation,

$$G_1(s) = \frac{r(s)}{\alpha(s)} = -\frac{mg}{\left(\frac{J_b}{R^2} + m\right)s^2} \tag{8.18}$$

And then set up the angle model of the ball – beam system. According to the geometry relation, there is equation (8.19), which is shown the relation of gear angle θ with beam angle α:

$$\sin\alpha = \frac{\sqrt{l^2 - [(d - d\cos\theta) - (L - L\cos\alpha)]^2} - (l - d\sin\theta)}{L} \tag{8.19}$$

Obviously, the relation of two angle is nonlinear. By means of MATLAB simulation, draw the relation curve, it is shown in Figure 8.6. It is clear from Figure 8.6, that the horizon position of beam, the angle of gear will keep in the ±0.5 radians, the two angle relation wshould be considered as linear.

这样由 $\frac{\mathrm{d}}{\mathrm{d}t}\left(\frac{\partial L}{\partial \dot{q}_i}\right) - \frac{\partial L}{\partial q_i} = \tau_i, i = 1, 2, \ldots, n$ 可以得到：

$$\begin{cases} \left(\frac{J_b}{R^2} + m\right)\ddot{r} - mr\dot{\alpha}^2 + mg\sin\alpha = 0 \\ (J_l + mr^2)\ddot{\alpha} + 2mr\dot{r}\dot{\alpha} + \left(mgr + \frac{1}{2}MgL\right)\cos\alpha = \tau \end{cases} \tag{8.16}$$

我们仅需要小球位置与横杆角度之间的数学关系，因此我们取式（8.16）的第一式即可。因为我们期望角度 α 在 0 附近，因此我们可以在 0 附近对其进行线性化。得到式（8.17）：

$$\ddot{r} = \frac{mg\alpha}{\frac{J_b}{R^2} + m} \tag{8.17}$$

取拉普拉斯变换可得：

$$G_1(s) = \frac{r(s)}{\alpha(s)} = -\frac{mg}{\left(\frac{J_b}{R^2} + m\right)s^2} \tag{8.18}$$

其次建立球杆系统的角度模型。由基本几何关系公式可以推出，齿轮转角 θ 和球杆转角 α 之间的关系式为（8.19）所示：

$$\sin\alpha = \frac{\sqrt{l^2 - [(d - d\cos\theta) - (L - L\cos\alpha)]^2} - (l - d\sin\theta)}{L} \tag{8.19}$$

很明显，二者之间为非线性关系。在这里用 MATLAB 进行仿真，绘制二者的关系曲线。如图 8.6 所示。通过仿真图可以看出，在横杆的水平位置，齿轮转角在 ±0.5 弧度的范围内，可以将其近似为线性关系。

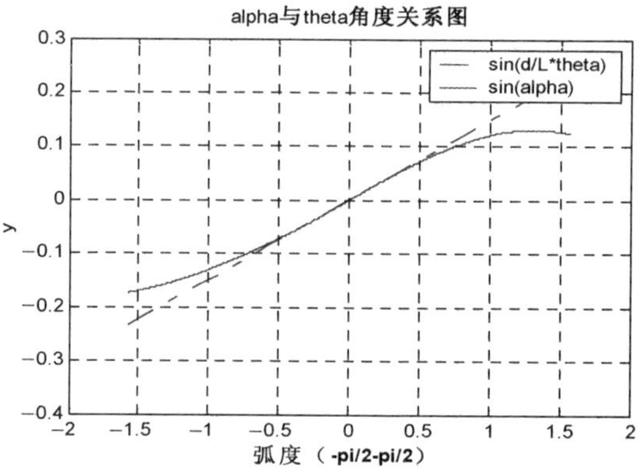

Figure 8.6 The Relation Curve Between Gear Angle with Beam Rotation Angle

图 8.6 齿轮转角与横杆转角关系图

$$\alpha = \frac{d}{L}\theta \tag{8.20}$$

Take the Laplace transformation:

$$G_2(s) = \frac{\alpha(s)}{\theta(s)} = \frac{d}{L} \tag{8.21}$$

The ball - beam will be combined as two series system: Mechanism part and angle translation part. It is shown as Figure 8.7.

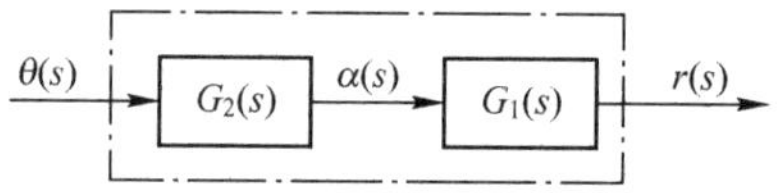

Figure 8.7 Block Diagram of Ball - beam System

图 8.7 球杆系统方框图

So, the open loop transfer function of Ball - beam system is as following:

$$G(s) = \frac{r(s)}{\theta(s)} = G_2(s)G_1(s)$$
$$= -\frac{mgd}{L\left(\frac{J_b}{R^2} + m\right)}\frac{1}{s^2} = \frac{c}{s^2} \tag{8.22}$$

with the relevant numerical value substituted, we have a new transfer function:

$$G(s) = \frac{0.7}{s^2} \tag{8.23}$$

Formula (8.23) is the transfer function of Ball - beam system.

$$\alpha = \frac{d}{L}\theta \tag{8.20}$$

取拉普拉斯变换可得：

$$G_2(s) = \frac{\alpha(s)}{\theta(s)} = \frac{d}{L} \tag{8.21}$$

球杆系统可看成两个子系统串联而成：球杆机械部分、角度转换部分。如 8.7 所示。

因此整个球杆系统的开环传递函数为：

$$G(s) = \frac{r(s)}{\theta(s)} = G_2(s)G_1(s)$$
$$= -\frac{mgd}{L\left(\frac{J_b}{R^2} + m\right)}\frac{1}{s^2} = \frac{c}{s^2} \tag{8.22}$$

代入各数值，可得传递函数：

$$G(s) = \frac{0.7}{s^2} \tag{8.23}$$

式（8.23）就是球杆系统的传递函数。

8.2.4 Ball - beam System Modelling Analysis

According to the modern control theory, the state space equation of control system will be describe as following:

$$\begin{aligned} \dot{X} &= AX + Bu \\ Y &= CX + Du \end{aligned} \tag{8.24}$$

Transfer the ball - beam system to state space model, assume $x_1 = r, x_2 = \dot{r}, u = \theta$, we get equation (8.25)

$$\begin{cases} \begin{pmatrix} \dot{x}_1 \\ \dot{x}_2 \end{pmatrix} = \begin{pmatrix} 0 & 1 \\ 0 & 0 \end{pmatrix} \begin{pmatrix} x_1 \\ x_2 \end{pmatrix} + \begin{pmatrix} 0 \\ 0.7 \end{pmatrix} u \\ y = x_1 \end{cases} \tag{8.25}$$

It is obviously from formula (8.25), the state space formula is a controllable type, that means the ball - beam system is controllable.

The stability of ball - beam system will be analyzed, by MATLAB, to simulate the open response for ball - beam system. write and run a step. m file a unit step response curve will be shown in Figure 8. 8.

According to the open loop curve, the step response curve is divergence, so the open loop of ball - beam system is unsteady. So insert a suitable controller is necessary and it will realize the steady control for ball - beam system.

8.2.4 球杆系统模型分析

由现代控制理论原理可知，控制系统的状态空间方程可写成如下形式：

$$\begin{aligned} \dot{X} &= AX + Bu \\ Y &= CX + Du \end{aligned} \tag{8.24}$$

将球杆系统的模型转化为状态空间模型，设 $x_1 = r, x_2 = \dot{r}, u = \theta$，则可以得到式（8.25）的状态空间：

$$\begin{cases} \begin{pmatrix} \dot{x}_1 \\ \dot{x}_2 \end{pmatrix} = \begin{pmatrix} 0 & 1 \\ 0 & 0 \end{pmatrix} \begin{pmatrix} x_1 \\ x_2 \end{pmatrix} + \begin{pmatrix} 0 \\ 0.7 \end{pmatrix} u \\ y = x_1 \end{cases} \tag{8.25}$$

由式（8.25）我们可以看出，状态空间方程为能控标准型，球杆系统是可控的。

再分析球杆系统的开环稳定性。在这里借助 MATLAB 软件对球杆系统开环响应进行仿真。编程实现一个 step. m 文件。执行这个文件，得到给定输入为一个 1 的阶跃信号时系统的响应曲线，如图 8. 8 所示。

由开环响应曲线可以看出，开环阶跃响应曲线是发散的，因而系统开环不稳定，必须添加适当的控制器才能实现球杆系统的稳定控制。

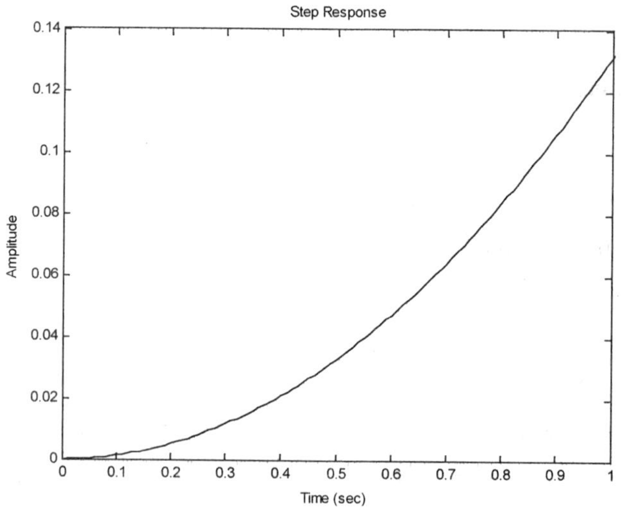

Figure 8. 8 The Step Response Curve of Open Loop System of Ball - beam

图 8. 8 球杆系统开环阶跃响应曲线

8. 3 Fuzzy Controller Design

8. 3. 1 The Component and Principle of Fuzzy Controller

The basic idea of fuzzy control is to use the machine to simulate human beings action. The block diagram of fuzzy control system is shown in Figure 8. 9. It consists of input, fuzzy controller, output, actuator, and plant. From Figure 8. 9, it is obvious that the system is feedback control system, but the controller is realized by fuzzy controller.

So the key component is fuzzy controller. The principle diagram of fuzzy controller is shown as Figure 8. 10.

According to the Figure 8. 10, there are 4 parts in a fuzzy controller: Fuzzy interface, knowledge base, inference engine, and defuzzification interface.

8. 3 模糊控制器设计

8. 3. 1 模糊控制器组成及原理

模糊控制的基本思想是用机器模拟人对系统的控制。模糊控制系统基本结构如图 8. 9 所示：由输入通道、模糊控制器、输出通道、执行机构与被控对象组成。由图 8. 9 可以看出，系统为常规闭环控制系统，控制器部分用模糊控制器实现。

因此，整个模糊控制系统的核心部分就是模糊控制器。其组成结构如图 8. 10 所示。

由图 8. 10 可知。模糊控制器包括 4 个部分，模糊化接口、知识库、推理机、解模糊接口。

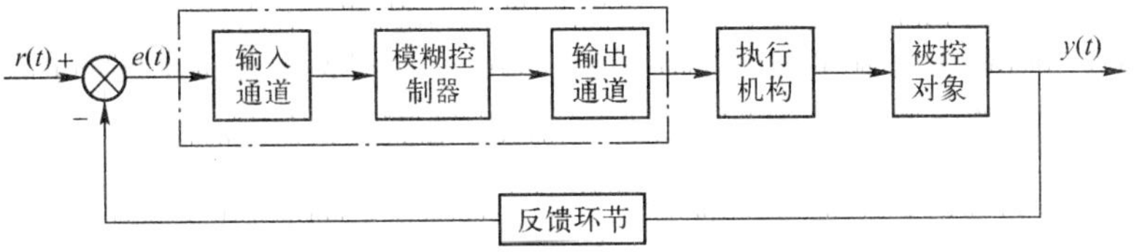

Figure 8.9 Structure Diagram of Fuzzy Control System

图 8.9 模糊控制系统基本结构图

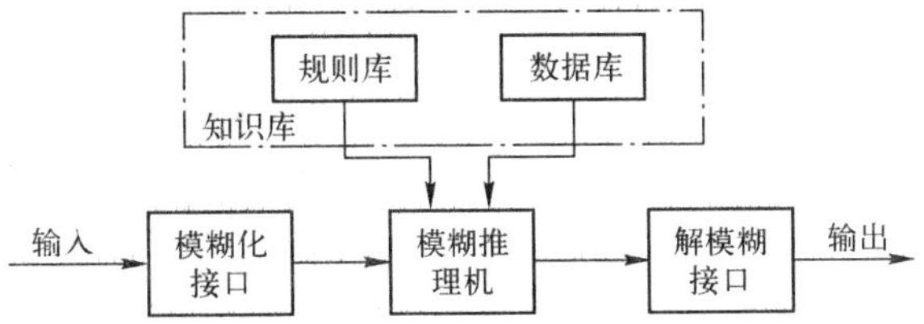

Figure 8.10 Diagram of Fuzzy Controller

图 8.10 模糊控制器结构图

In fuzzy control system, the general control steps are: the computer keeps sampling continuously to get the value of controlled variable, and compares the sampling value with the given input, we will get the error "e" . Let error "e" as an input of the fuzzy controller, and fuzzy e is obtained by fuzzification of the error, so the fuzzy e will be described by fuzzy language.

在模糊控制系统中，其一般控制步骤为：计算机不断采样获取被控制量的采样值，然后将采样值与给定值进行比较得到误差 e。选取误差 e 作为模糊控制器的一个输入，对误差 e 进行模糊化得到其模糊量，该模糊量可用相应的模糊语言表示。

After we get the subset E from fuzzy language sets of error "e", E and fuzzy control rule R make fuzzy decision according to the conjunction rule of inference. It is shown in fornala (8.26) that the result U is fuzzy control vector:

得到误差 e 的模糊语言集合的一个子集 E，根据推理的合成规则 E 与模糊控制规则 R 进行模糊决策，得到模糊控制向量 U 如式（8.26）所示。

$$U = E \circ R \tag{8.26}$$

$$U = E \circ R \quad (8.26)$$

And then the fuzzy vector U is transformed into the precise quantity, after the D – A transformation, it will be transformed as analog quantity to the actuator. The plant will be controlled by actuator.

将模糊向量 U 解模糊转化为精确量，经 D – A 转换为模拟量给执行机构，从而控制被控对象。

8.3.2 The Design and Experiment of Fuzzy Controller for Ball – beam System

8.3.2.1 The Content of Fuzzy Controller

There are no fixed modelling method for Fuzzy controller design. Mostly, there are five steps in the controller design:

1) Select structure in fuzzy controller;

2) Select linguistic variable, fuzzification for the variable, and confirm the membership function;

3) Design the rule of fuzzy and operation;

4) Inquire the fuzzy output and difuzzification;

5) Select parameters about fuzzy controller;

According to the 5 steps, we can design the fuzzy controller. to ball – beam system.

8.3.2.2 The Design of Fuzzy Controller

(1) The Selection of Fuzzy Controller Structure

The ball – beam system has nonlinear character. In order to get satisfied control performance, we not only need to observe the ball position error "e", but also to observe the error variation "ec", that is, the speed of the ball. The only one control variable is the gear rotation angle θ. So, the fuzzy controller is selected as two input and one output in two dimensional structure. It is shown in Figure 8.11. Where the K_e, K_{ec} is the quantitative factor, K_y is the scale factor.

8.3.2 球杆系统模糊控制器设计与实验

8.3.2.1 模糊控制器设计内容

对于模糊控制器的设计，并没有固定的方法。但一般来说，模糊控制器的设计过程主要包括以下五个步骤。

1）模糊控制器结构的选择；

2）选取语言变量，进行模糊化，对隶属度函数进行确定；

3）设计模糊控制规则及运算合成；

4）查询模糊输出与解模糊；

5）模糊控制器控制参数的选择。

根据以上 5 个设计步骤，对球杆系统模糊控制器进行设计。

8.3.2.2 球杆系统模糊控制器设计

（1）球杆系统模糊控制器结构选择

由于球杆系统具有非线性，为了得到良好的控制性能，不仅需要观测小球位置误差 e，还要观测其误差变化 ec，也就是小球的速度。控制量只有一个，就是齿轮旋转角度 θ。因此模糊控制器选择两输入 - 单输出二维结构。球杆系统控制结构如图 8.11 所示。其中 K_e，K_{ec}为量化因子，K_y 为比例因子。

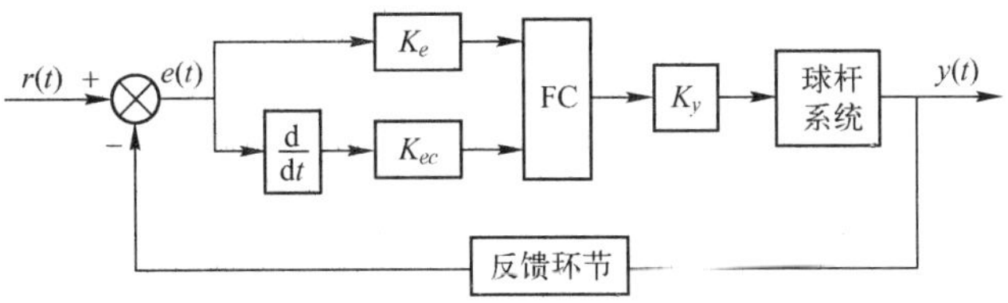

Figure 8. 11 The Block Diagram of Ball – beam Fuzzy Control System

图 8. 11 球杆系统模糊控制结构图

(2) Determination of linguistic variables and membership functions

Because the system structure is two input – single output structure, so the input and output variables have a total of three, the input variables are the ball position error "e", the ball position error rate of change, that is, the speed of the ball "ec". Output variable is output control angle θ. Assume that the error linguistic variable is E, the error rate linguistic variable is EC, and the output of the system linguistic variable is U. The actual sampling position of the Ball – beam system is r, the reference position is r_d.

Select:

$$e = r - r_d \tag{8.27}$$

$$ec = (r_2 - r_d) - (r_1 - r_d) = r_2 - r_1 \tag{8.28}$$

For the error variable E, the universe is {−6, −5, −4, −3, −2, −1, 0, 1, 2, 3, 4, 5, 6}, linguistic value is {negative big (NB), negative middle (NM), negative small (NS), zero (Z), positive small (PS), positive middle (PM), positive big (PB)} Respectively, the current position of the ball r relative to the set value r_d is: "smallest", "smaller", "small", "just", "large", "larger", "largest".

（2）确定语言变量及隶属度函数

由于系统结构为两输入 – 单输出结构，因此输入输出变量共有三个，输入变量分别为小球位置误差 e，小球位置误差变化率——即小球的速度 ec。输出变量为输出控制角度 θ。取误差语言变量为 E，误差变化率语言变量为 EC，系统输出控制量为 U。设球杆系统实际采样位置为 r，参考位置为 r_d。

取：

$$e = r - r_d \tag{8.27}$$

$$ec = (r_2 - r_d) - (r_1 - r_d) = r_2 - r_1 \tag{8.28}$$

对于误差变量 E，取论域为 {−6, −5, −4, −3, −2, −1, 0, 1, 2, 3, 4, 5, 6}，语言值为：{负大（NB），负中（NM），负小（NS），零（Z），正小（PS），正中（PM），正大（PB）}。分别表示当前小球位置 r 相对于设定值 r_d 为："极小"、"很小"、"偏小"、"正好"、"偏大"、"很大"、"极大"。

For the error rate of EC, the universe is {-6, -5, -4, -3, -2, -1, 0, 1, 2, 3, 4, 5, 6}, linguistic value: {negative big (NB), negative middle (NM), negative small (NS), zero (Z), positive small (PS), positive middle (PM), positive big (PB)}. The current speed of the ball: "left roll fastest", "left roll faster", "left roll fast", "ball stop", "right roll fast", "right roll faster", "right roll fastest".

对于误差变化率EC，取论域为{-6，-5，-4，-3，-2，-1，0，1，2，3，4，5，6}，语言值为：{负大(NB)，负中(NM)，负小(NS)，零(Z)，正小(PS)，正中(PM)，正大(PB)}。分别表示当前小球的速度“左滚极快”、“左滚很快”、“左滚偏快”、“小球静止”、“右滚偏快”、“右滚很快”、“右滚极快”。

For the control value U, the universe is selected as {-6, -5, -4, -3, -2, -1, 0, 1, 2, 3, 4, 5, 6}, linguistic value is {negative big (NB), negative middle (NM), negative small (NS), zero (Z), positive small (PS), positive middle (PM), positive big (PB)}. It means that the current gear angle are "up - deviate biggest", "up - deviate bigger", "up - deviate big", "horizon is zero", "under - deviate big", "under - deviate bigger", “under - deviate biggest”.

对于输出控制量为U，取论域为{-6，-5，-4，-3，-2，-1，0，1，2，3，4，5，6}，语言值为：{负大(NB)，负中(NM)，负小(NS)，零(Z)，正小(PS)，正中(PM)，正大(PB)}。分别表示当前齿轮偏角“上偏极大”、“上偏很大”、“上偏偏大”、“水平为零”、“下偏偏大”、“下偏很大”、“下偏极大”。

The three membership functions are triangle membership functions; the values of membership functions are shown in Figure 8.12. The values in Figure 8.12 are quantified according to E, EC, U, and shown in table 8.2. First column are linguistic values and first raw are true numerical values. In the middle of the table 8.2, are the membership of linguistic values to accurate value.

三者隶属度函数均选择三角形隶属度函数。具体隶属函数值如图8.12所示。将图8.12根据E、EC、U量化，量化后的E、EC、U的隶属度函数表见表8.2。第一列表示各个语言值，第一行表示具体的数值。中间即为各语言值在精确量上的隶属度。

(3) Set up the fuzzy rule

The selection of fuzzy control variable follows such rules: when the error is biggest or bigger, the choice of fuzzy control variable should be based on the elimination of error. When the error is small, the choice of fuzzy control is based on the stability of the system, mostly to prevent the heavy percent overshoot.

(3) 建立模糊控制规则

模糊控制量的选取遵循以下规则：当误差极大或很大时，模糊控制量的选择应以消除误差为主。当误差较小时，模糊控制量的选择就以系统的稳定性为主，防止系统超调。

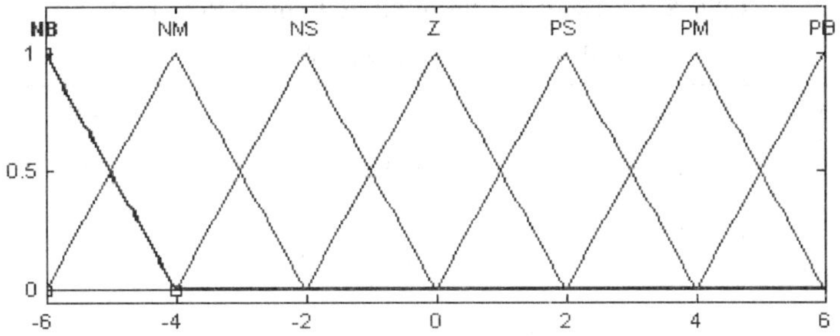

Figure 8. 12 The Membership of E, EU, U

图 8. 12 E、EC、U 隶属函数图

Table 8. 2 The Membership Table of E, EU, U

表 8. 2 E、EC、U 隶属函数表

	-6	-5	-4	-3	-2	-1	0	1	2	3	4	5	6
PB	0	0	0	0	0	0	0	0	0	0	0	0. 5	1
PM	0	0	0	0	0	0	0	0	0	0. 5	1	0. 5	0
PS	0	0	0	0	0	0	0	0. 5	1	0. 5	0	0	0
Z	0	0	0	0	0	0. 5	1	0. 5	0	0	0	0	0
NS	0	0	0	0. 5	1	0. 5	0	0	0	0	0	0	0
NM	0	0. 5	1	0. 5	0	0	0	0	0	0	0	0	0
NB	1	0. 5	0	0	0	0	0	0	0	0	0	0	0

According to the experience of the ball - beam control system, the fuzzy control rule is set up as table 8. 3:

根据球杆系统控制经验，建立模糊控制规则表见表 8. 3。

First row is linguistic value of *E*, and first column is the linguistic value of *EC*.

第一行为 *E* 的语言值，第一列为 *EC* 的语言值。

(4) Fuzzy control table to ball - beam system fuzzy controller

(4) 计算球杆系统模糊控制器的模糊控制表

The fuzzy control rule table is calculated by using fuzzy inference rules. By means of Mamdani inference method, to adopt the Weighted Average Mean method, and the fuzzy control table is shown in table 8. 4.

应用模糊推理规则，计算模糊控制规则表。采用 Mamdani 推理方式，解模糊采用加权平均法，经过计算得到的模糊控制表见表 8. 4，其中第一行为误差变量 *E* 的取值，第一列为误差变化率 *EC* 的取值。

(5) The selection of quantization factor and scaling

(5) 量化因子与比例因子的选择

Table 8.3 The Ball – Beam Fuzzy Control Rule Table

表 8.3 球杆系统模糊控制规则表

U	NB	NM	NS	Z	PS	PM	PB
NB	PB	PB	PM	PM	PS	Z	Z
NM	PB	PM	PS	PS	PS	Z	Z
NS	PM	PM	PS	PS	Z	NS	NS
Z	PM	PS	PS	Z	NS	NS	NM
PS	PS	PS	Z	NS	NS	NM	NM
PM	Z	Z	NS	NS	NS	NM	NB
PB	Z	Z	NS	NM	NM	NB	NB

Table 8.4 The Ball – Bean System Fuzzy Control Table

表 8.4 球杆系统模糊控制表

	-6	-5	-4	-3	-2	-1	0	1	2	3	4	5	6
-6	5.4	5.4	5.4	4.3	4	4	4	3	2	1	0	0	0
-5	5.4	4.3	4.3	4.3	4	3	3	3	2	1	0	0	0
-4	5.4	4.3	4	4	4	3	2	2	2	1	0	0	0
-3	4.3	4.3	4	3	3	3	2	1	1	0	-1	-1	-1
-2	4	4	4	3	2	2	2	1	0	-1	-2	-2	-2
-1	4	3	3	3	2	1	1	0	-1	-1	-2	-3	-3
0	4	3	2	2	2	1	0	-1	-2	-2	-2	-3	-4
1	3	3	2	1	1	0	-1	-1	-2	-3	-3	-3	-4
2	2	2	2	1	0	-1	-2	-2	-2	-3	-4	-4	-4
3	1	1	1	0	-1	-1	-2	-3	-3	-3	-4	-4.3	-4.3
4	0	0	0	-1	-2	-2	-2	-3	-4	-4	-4	-4.3	-5.4
5	0	0	0	-1	-2	-3	-3	-3	-4	-4.3	-4.3	-4.3	-5.4
6	0	0	0	-1	-2	-3	-4	-4	-4	-4.3	-4.3	-5.4	-5.4

According to the influence of quantization factor and scaling factor on system performance, and the ball – beam structure parameters, the original parameters is chosen as $K_e = 25$, $K_{ec} = 20$, $K_y = 0.5$.

根据量化因子和比例因子对系统性能的影响，结合球杆系统的具体结构参数，初始参数选择 $K_e = 25$，$K_{ec} = 20$，$K_y = 0.5$。

The ball – beam fuzzy controller is completed.

球杆系统模糊控制器设计完成。

8.3.3 The Realize of Ball-beam System Fuzzy Controller

After the fuzzy controller is designed, the fuzzy control algorithm is simulated by MATLAB. MATLAB fuzzy toolbox is used to establish the fuzzy control FIS file of ball – beam system, the language variable, the membership function and the control rules are set up in this FIS file. The Simulink simulation model is shown as Figure 8.13, K_e, K_{ec} are the error quantization factor and error variable rate factor. K_u is the control scale factor; it will be adjusted with the simulation. The fuzzy control algorithm simulation curve is shown as Figure 8.15. As a contrast, build the PID control simulation block diagram as shown in figure 8.14. The PID control algorithm response curve is shown as Figure 8.16.

8.3.3 球杆系统模糊控制器实现

设计好模糊控制器，运用 MATLAB 对模糊控制算法进行仿真。利用 MATLAB 模糊工具箱建立球杆系统模糊控制 FIS 文件，在此文件中设置语言变量、隶属度函数及控制规则。搭建模糊控制 Simulink 仿真模型，如图 8.13 所示，K_e、K_{ec}分别为误差量化因子和误差变化量化因子，K_u 为控制量比例因子，具体数值仿真时有所调整。模糊控制算法仿真响应曲线如图 8.15 所示。作为对比，同时搭建 PID 控制仿真框图，如图 8.14 所示。PID 控制算法响应曲线如图 8.16 所示。

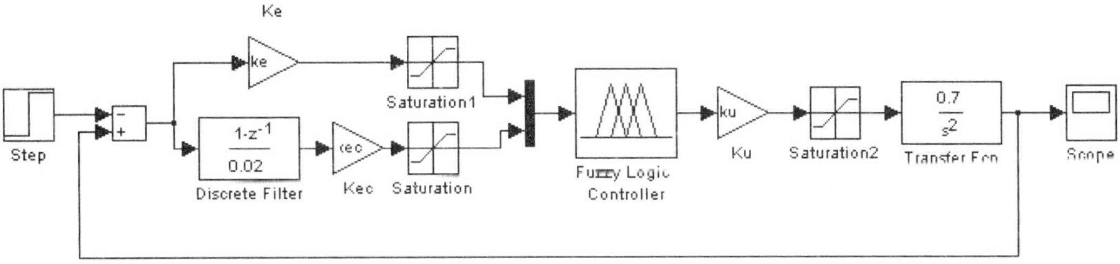

Figure 8.13 The Ball – Beam Fuzzy Control Simulation Block Diagram

图 8.13 球杆系统模糊控制仿真框图

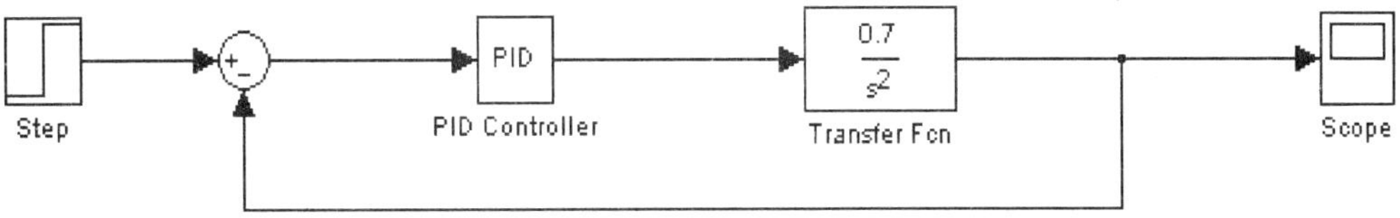

Figure 8.14 The Ball – beam PID Control Simulation Block Diagram

图 8.14 球杆系统 PID 控制仿真框图

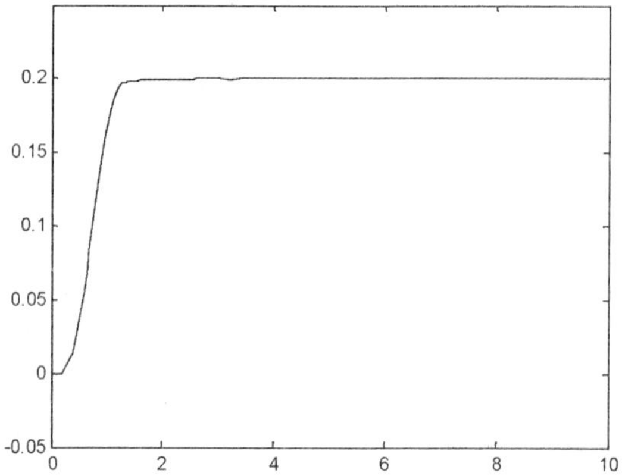

Figure 8. 15 The Response Curve of Ball – beam Fuzzy Control Simulation

图 8. 15 球杆系统模糊控制仿真响应曲线

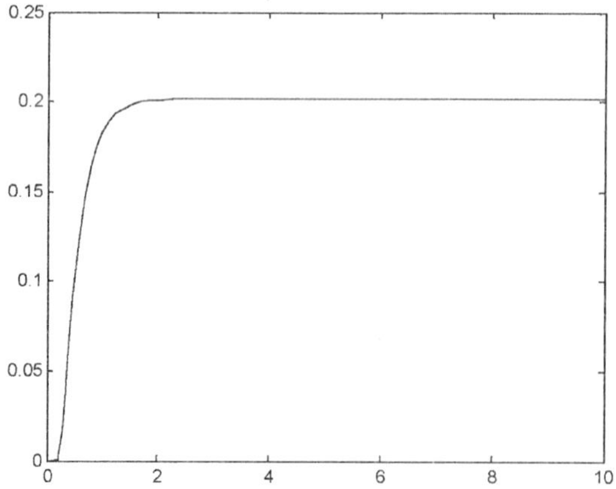

Figure 8. 16 The Ball – beam PID Control Simulation Response Curve

图 8. 16 球杆系统 PID 控制仿真响应曲线

According to figure 8. 15 and figure 8. 16, the simulation results show that, in theory, the two algorithm have similar control performance, but for the nonlinear control system, the fuzzy control algorithm has better control performance comparing with the PID control algorithm. Even more, the parameter adjustment of fuzzy controller is simpler than PID control algorithm.

由图 8. 15 和图 8. 16 仿真图可以看出，理论上二者有着相似的控制性能，但对于非线性被控制系统，模糊控制算法理论上有着比 PID 控制算法更好的控制性能。而且模糊控制器的参数调整要比 PID 控制算法简便。

And then, the fuzzy control algorithm is applied to the actual ball – beam control system, to adjust the parameters of the system repeatedly. The system is stable but has big steady state error sometimes. when the beam in different position, the steady state error will be changed also.

将上述模糊控制算法用于球杆系统实际控制，反复调整系统的参数。系统能够稳定，但有时会有较大的静差。并且球杆位置摆放不同，系统静差也会有所变化。

For the control of these problems, the reasons are analyzed as follows:

针对控制中出现的这些问题，分析原因如下：

1) The system sample error. the sampling time of the ball – beam system is 0.02s, and the sampling noise will affect the system.

1）系统采样误差，球杆系统的采样时间为0.02s，采样时噪声会对系统有所影响。

2) The accuracy of the ball – beam system itself. The ideal control tgoalet is when the gear angle is zero, the system is stable, but in fact, the gear level angle is not zero when the system is stable, and the desktop is not completely horizontal, so it will lead to error in gear angle. This will affect the final control precision of the system, so it is necessary to eliminate this effect to achieve better control performance.

2）球杆系统本身的精度问题。理想的控制目标是齿轮偏角为零时，系统刚好稳定，但实际上，球杆系统稳定时齿轮水平偏角不为零，再加上摆放的桌面也不完全水平，因此，系统本身的齿轮偏角有一定的误差。这会对系统最终的控制精度有所影响，必须消除此种影响才能实现较好的控制性能。

3) The characteristics of conventional fuzzy controller. In the general fuzzy control system, the fuzzy controller is realized simple and fast. Generally, the structure of the two – dimensional fuzzy controller is adopted, the input variables are the system error E and the error variation rate EC, so it has the similar performance to the conventional PD controller. According to the linear control theory, the controller can obtain good dynamic characteristics, but it cannot eliminate the static error of the system, to realize the no steady state error system.

3）常规模糊控制器本身的特点。在一般模糊控制系统中，考虑模糊控制器实现的简便性和快速性，通常采用二维模糊控制器结构，取输入变量为系统误差E和误差变化EC，因此它具有类似常规PD控制器的性能。由线性控制理论可知，采用该控制器可以获得良好的动态特性，但无法消除系统的静态误差，实现无差系统。

Based on the above factors, in order to improve the control performance of the fuzzy controller and eliminate the static error, the integral item will be added to fuzzy controller. The improved fuzzy controller will be introduced in the next section.

综合以上因素，为了改善模糊控制器的控制性能，消除静态误差，在模糊控制器上引入模积分因子。下边将对改进模糊控制器进行介绍。

8.3.4 Improvement and Simulation of The Ball - beam Fuzzy Controller

8.3.4 球杆系统模糊控制器改进与仿真

The characteristic of the conventional fuzzy controller determines that it cannot eliminate the static error of the system. Why? Analyz one-dimensional fuzzy controller as an example. Assume the fuzzy subset of input variable e in universe E is $B_i(i=1,2,\cdots,m)$; The fuzzy subset of the output variable u in the universe U is $A_i(i=1,2,\cdots,n)$. Through the control function of the fuzzy controller, when the system is perturbed with ε, the error is still in the desired fuzzy subset B^*. If the fuzzy control rule based on Expert System is as following:

if $E=B^*$ then $U=A^*$,

if $E=B_i$ then $U=A_i$.

Assuming the original condition is $e=B^*$, $u=A^*$, the system is stable. When there is an external disturbance to get in, the system output will deviate from a given reference value, that means $e=B_i$, and the output of system is A_i, and select $u=A_i$, in order to counteract the influence of disturbance, and make the system will back to the original value. But when $e=B^*$, the fuzzy controller becomes $u=A^*$ immediately. The balance state is broken out, the error will become $e=B_i$ again, repeatedly. The phenomena of limit cycle oscillation will appear in the ball - beam control system. The limit cycle phenomenon, makes the error value fall in between B^* and B_i in a fuzzy

常规模糊控制器本身的特点，决定了它不能消除系统静差。分析其原因，以简单一维模糊控制器为例。设输入变量 e 在论域 E 上的模糊子集为 $B_i(i=1,2,\cdots,m)$；输出变量 u 在论域 U 上的模糊子集为 $A_i(i=1,2,\cdots,n)$。通过模糊控制器的控制作用，使得系统有扰动量 ε 时，其误差仍在期望的模糊子集 B^* 中。如果基于专家知识的模糊控制规则为：

如果 $E=B^*$，那么 $U=A^*$，

如果 $E=B_i$，那么 $U=A_i$。

假设初始条件为 $e=B^*$，$u=A^*$ 时，系统处于稳定状态；在外界扰动作用下，系统输出偏离给定值，使误差 $e=B_i$，此时系统输出为 A_i，取 $u=A_i$，以此抵消扰动的影响，使系统输出回到原来的值。但一旦当 $e=B^*$ 时，模糊控制器马上又变成 $u=A^*$；刚刚达到的平衡状态又将不复存在，偏差又将变成 $e=B_i$，依次往复，系统将产生极限环振荡现象。极限环现象的存在，使系统的误差终值落在 B^* 和 B_i 之间的某一模糊子集内，而控制的终值为相应的 A^* 和 A_i 之间的模糊子集。因为系统存在静差，需要在常规模糊控制器基础之上引入模糊积分环节才能消除这种静差。

subset, the final control value will fall in the interval of A^* and A_i, which is a fuzzy subset. Because the system has steady state error, so it is necessary to add fuzzy integral component to the conventional fuzzy controller to eliminate the static error.

About the fuzzy integral component, a solution using the structure of fuzzy controller is shown as figure 8.17. It consists of one conventional integral controller and one two - dimension fuzzy controller with parallel connection. The output of *PI* controller is $u_i = K_I \sum_i e_i$, which plus the two - dimensional fuzzy controller of u_f, we set the total output of *PID* controller. Here $e(t)$ keeps vary continuously, so that the $u_i = K_I \sum_i e_i$ is a continuous variable also.

对于模糊积分环节，一种解决方案采用如图 8.17 模糊控制器结构。它是由一个常规积分控制器和一个二维模糊控制器并联而成的。*PI* 控制器的输出为 $u_i = K_I \sum_i e_i$ 和二维模糊控制器的控制量 u_f 相加，作为模糊 *PID* 控制器的总输出。这里 $e(t)$ 是连续变化的，因此 $u_i = K_I \sum_i e_i$ 也是一个连续量。

This kind of Fuzzy *PID* controller can not only eliminate the limit oscillation, but also eliminate the static error of the system. The system will become a no steady state error fuzzy control system.

这种模糊 *PID* 控制器不仅可以消除极限振荡，而且可完全消除系统的静差，使系统成为无差模糊控制系统。

In addition to the above design scheme, there is a fuzzy *PID* controller which integrates the fuzzy value of the error *e*. The structure is shown in Figure 8.18.

除上述方案外，还有一种对误差 *e* 的模糊值进行积分的模糊 *PID* 控制器方案。其结构图如图 8.18 所示。

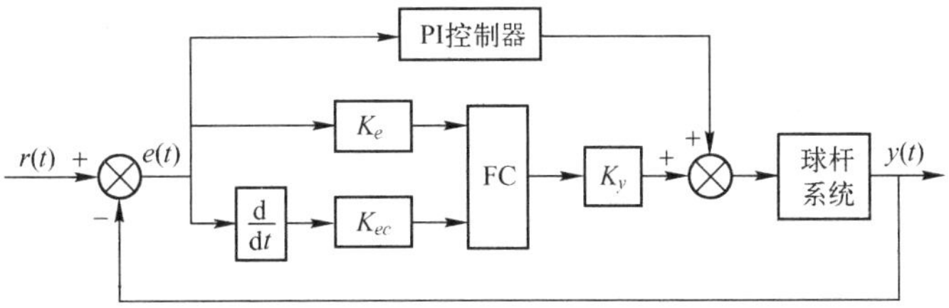

Figure 8.17 The Ball - beam Fuzzy Controller Block Diagram

图 8.17 球杆系统模糊 PID 控制器框图

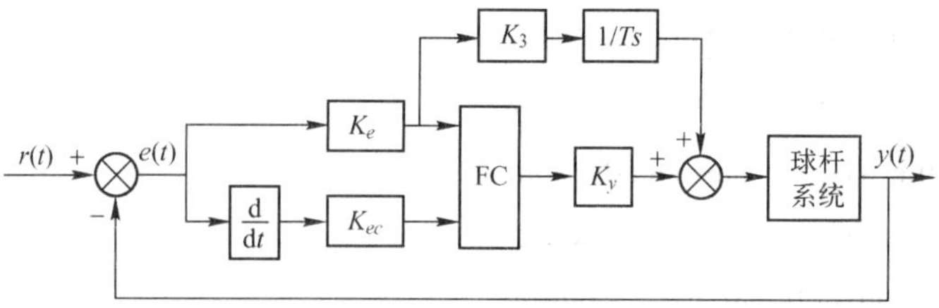

Figure 8. 18 The Ball – beam Fuzzy PID Controller Block Diagram for Error *e*

图 8. 18 对误差 *e* 的模糊值积分的球杆系统模糊 PID 控制器框图

The fuzzy *PID* controller which integrates the fuzzy value of the error *e* can be used to eliminate the large static error of the system, however, it is necessary to eliminate the limit cycle oscillation near the zero point, which leads to an increase in the number of fuzzy rules, the complexity of fuzzy controller design is increased accordingly.

这种对误差 e 的模糊值进行积分的模糊 *PID* 控制器，可以用来消除大的系统静差，但是要消除零点附近的极限环振荡会导致模糊规则数的增加，相应地增加了模糊控制器设计的复杂程度。

Improvement scheme: in order to eliminate the steady state error, fuzzy integral is added to the designed fuzzy controller. As for an integral coefficient is added, so the scaling factor and the quantization factor are adjusted accordingly. The new fuzzy integral controller is simulated by MATLAB software. The Simulink diagram is shown in Figure 8. 19: A *PID* controller is added, in which the scale factor and the differential system are zero, and only the integral item is added. The fuzzy controller still adopts in the former conventional fuzzy controller and the parameter setting is changed. The response curve of the system is shown as figure 8. 20, it is obviously that the performances are similar for conventional fuzzy controller and *PID* controller. But it will eliminate the static error than the general controller.

改进方案：为了消除静差，在原来已经设计好的模糊控制器基础之上引入模糊积分环节。相应地添加一个积分系数，比例因子和量化因子相应也要调整。利用 MATLAB 软件对新构建的模糊积分控制器进行仿真。其仿真框图如图 8. 19 所示：添加了一个 *PID* 控制器，其中比例系数和微分系统分别为零，仅添加积分作用。常规模糊控制器仍采用前面设计的模糊控制器，具体参数设置有所改变。系统响应曲线如图 8. 20 所示，从曲线可以算出，理论上的控制性能与常规模糊器和 *PID* 控制器相当。但与常规控制器相比，可以消除系统静态误差。

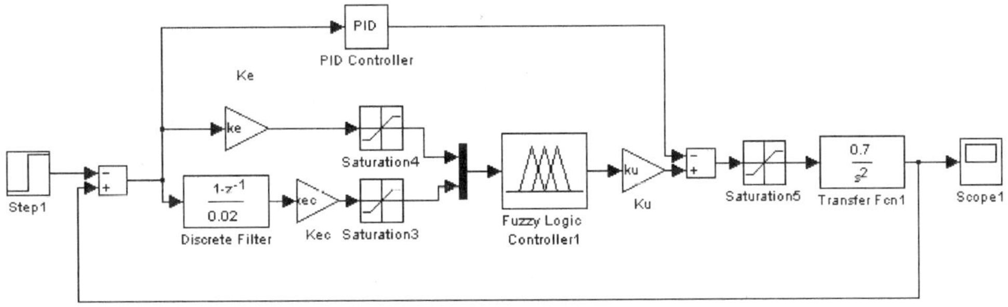

Figure 8.19 Simulink Diagram of Fuzzy PID Control for Ball - beam System

图 8.19 球杆系统模糊 PID 控制器仿真框图

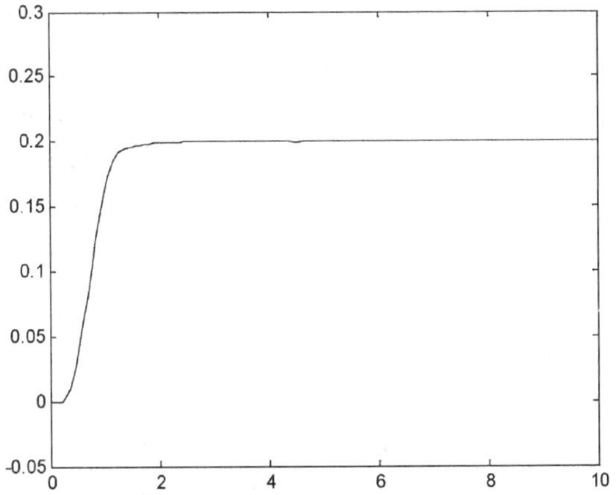

Figure8.20 Ball - beam System Fuzzy PID Controller Response Simulation Curve

图 8.20 球杆系统模糊 PID 控制器仿真响应曲线

8.4 The Physical Control Experiment of Ball - beam System

According to the simulation results in the above paragraph, PID controller, general fuzzy controller, fuzzy PID controller have similar control performance. These kind of controller will be used in the Ball - beam system very smoothly. The following control issue will be discussed.

MATLAB consists of two parts, one is fordata acquisition and processing, another is controller. The structure is shown in Figure 8.21.

8.4 球杆系统实际控制实验

由前面的仿真结果可能看出，PID 控制、常规模糊控制器、模糊 PID 控制器具有相似的控制性能，理论上都可以很好地控制球杆系统。下面就球杆系统实际控制问题进行讨论。

实际控制采用 MATLAB 软件。主要由两大部分组成，包括数据采集和处理、控制器部分。结构如图 8.21 所示：

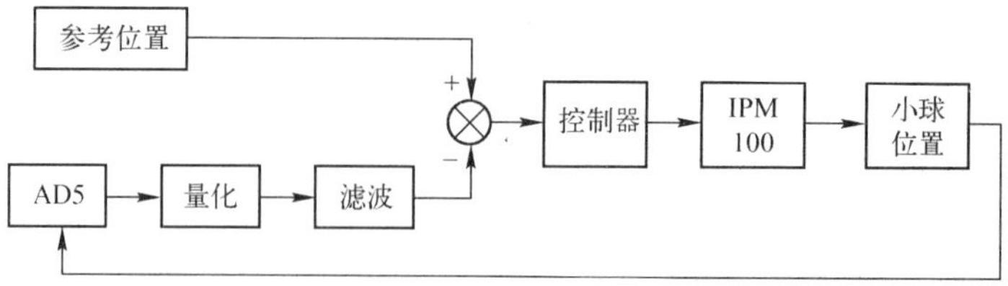

Figure 8. 21 The Block Diagram Of Physical Ball – beam System

图 8. 21 球杆系统实际控制框图

(1) Data acquisition and processing

This part gets the position of the ball by sampling, and then begins the data processing in order to obtain accurate data of the ball position. The ball position in the Ball – beam system is tested through the potentiometer output voltage. The potentiometer is connected with IPM100 intelligent controller AD conversion channel AD5, AD5 (16 bit) is in the range of 0 ~ 65535, corresponding to the voltage 0 ~ 5V, and corresponding to the position of the ball, 0 ~ 400mm. The process is: through the potentiometer voltage detection, the detected voltage is transferred to position signal, then gets the position signal in a filter to get rid of too big or too small data, so that the accurate position sample signal of ball will be obtained. Then the difference between the sampling data and the reference position is obtained, that means the error of position *e* will be got as the input of controller.

(2) Controller

According to position error information provided by the data sampling and processing, calculated by the designed control algorithm, sent the angle control law to IPM intelligent drive, the ball position will be controlled by IPM intelligent controller to adjust the angle of beam. This process is ongoing, sampling time is 0. 02s. Users can design their own controller by MATLAB software toolbox, to control system of Ball – beam.

(1) 数据采集和处理

此部分负责对小球的位置进行采样，并将采样数据进行处理，以获得小球位置的准确数据。球杆系统中小球的位置是通过电位计的输出电压来检测的。它和 IPM100 智能控制器的 AD 转换通道 AD5 相连，AD5 (16 位) 的范围为 0 ~ 65535，对应的电压为 0 ~ 5V，相应的小球位置为 0 ~ 400mm。具体过程为：通过电位器检测电压，将检测到的电压量化为位置信息，对量化的位置信息进行滤波，将过大或者过小的数据滤掉，得到小球的准确采样位置。然后对采样数据和参考位置求差，得到小球位置误差 *e*，作为控制器的输入。

(2) 控制器部分

根据数据采样和处理部分提供的位置误差信息，通过设计的控制算法进行计算，将得到的角度控制量送给 IPM 智能驱动器，由 IPM 智能控制器通过调整角度控制小球的位置。此过程是一直进行的，采样时间为 0. 02s。用户可以用 MATLAB 软件设计自己的控制器，对球杆系统进行控制。

The actual control of the *PID* algorithm is shown in Figure 8.22, and the actual control block diagram of the fuzzy integral is shown in figure 8.23. The response curve of *PID* is shown in Figurc 8.24, and the fuzzy integral response curve is shown in Figure 8.25.

PID 控制算法实际控制如图 8.22 所示，模糊积分实际控制框图如图 8.23 所示。*PID* 响应曲线如图 8.24 所示，模糊积分响应曲线如图 8.25 所示。

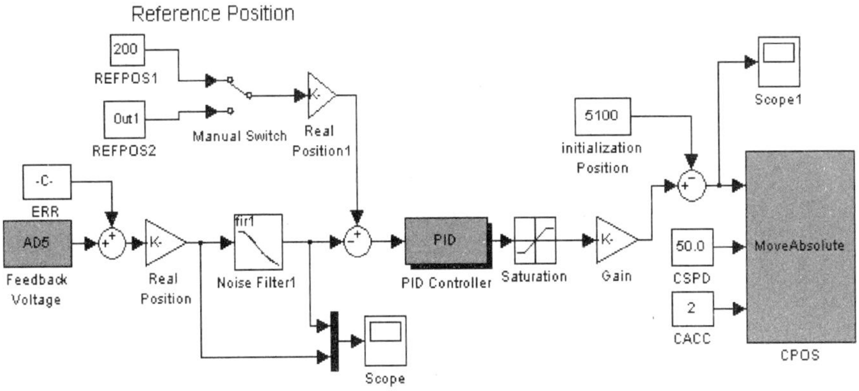

Figure 8.22 The Control Diagram of *PID* in Ball - beam System

图 8.22 球杆系统 *PID* 控制实际框图

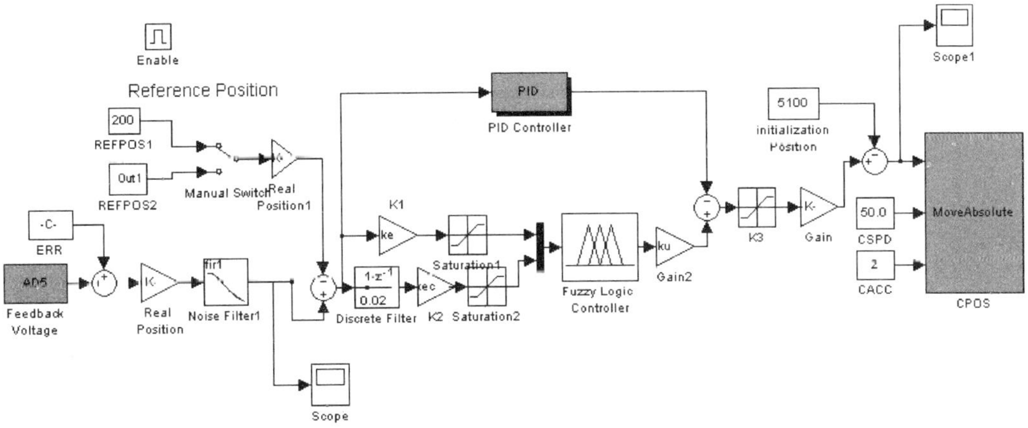

Figure 8.23 The Fuzzy Integration Control Diagram of Physical Ball - beam System

图 8.23 球杆系统模糊积分控制实际框图

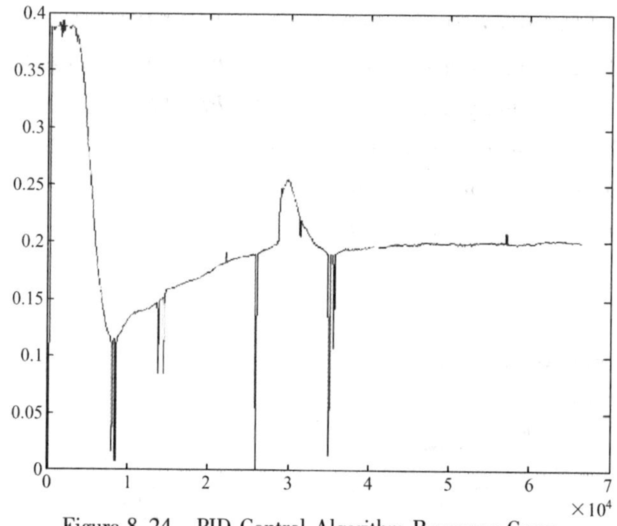

Figure 8.24 PID Control Algorithm Response Curve

图 8.24 PID 控制算法的响应曲线

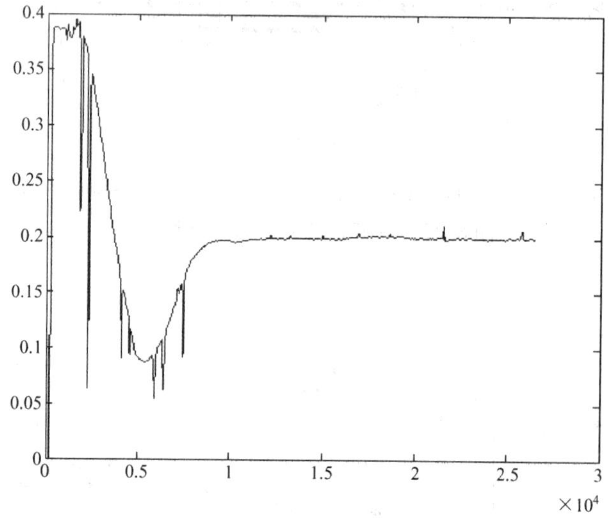

Figure 8.25 Fuzzy Integration Control Algorithm Response Curve

图 8.25 模糊积分控制算法响应曲线

Because the initial state of the ball - beam system is on the right side of the bar, the initial location is in 0.4 meters. From the actual response curve of Ball - beam system, we can see that the fuzzy integral controller in practical control system has a dynamic performance better than the PID controller. Even more, in terms of stability, fuzzy integral control algorithm and the PID algorithm can make the system stable and no static error.

因为球杆系统初始状态为小球在横杆的右侧，初始位置为 0.4 m。通过球杆系统实际响应曲线可以看出，模糊积分控制器在实际控制球杆系统时有着比 PID 控制器更好的动态性能。在稳定性方面，模糊积分控制算法和 PID 算法都 可使系统稳定且没有静差。

8.5 Summary

The composition and structure of the ball - beam system is introduced in this chapter firstly, and a brief description of its principle is given. According to the modeling of the ball - beam system, the Lagrange modelling method is selected. The ball - beam system is divided into three subsystems, and the mathematical model of the system is obtained. The open - loop transfer function of the whole system is obtained after Laplace transformation. It is obviously from the open - loop transfer function: ball - beam system is unstable, we must add a controller to make the system stable. This chapter focuses on the design of fuzzy control algorithm and fuzzy controller.

From the design of the fuzzy controller, the general design steps of the conventional fuzzy controller are described in this chapter, and the influence of the relevant parameters of the fuzzy controller on the system performance is explained in detail. According to the physical ball - beam system, a fuzzy controller is designed, including the design of language variable selection, membership function confirmation, fuzzy rules design. The simulation of fuzzy controller has been done and compareed with PID controller. In the application of the conventional fuzzy controller to the actual process, some problems were found, an improved scheme is used in the former controller and the results shown effective and reasonable of its control performance.

8.5 本章小结

本章首先介绍了球杆系统的组成及结构，并对其基本工作原理作了简要的描述。针对球杆系统建模，选择了拉格朗日建模方法。将球杆系统分为三个子系统进行了建模分析，得到了球杆系统的数学模型。对其进行变换，得到整个球杆系统的开环传递函数。由开环传递函数可以看出：球杆系统本身是不稳定的，必须添加合适的控制器才能使其稳定，本章针对模糊控制算法与模糊控制器的设计问题进行研究。

从模糊控制器的设计角度考虑，阐述了常规模糊控制器的一般设计步骤，并对模糊控制器的相关参数对系统性能的影响作了详细的说明。针对球杆系统实物模型设计了模糊控制器，包括语言量的选取、隶属度函数的确定、模糊规则的设计。对设计的模糊控制器进行了系统仿真并与 PID 控制器做了比较。在把常规模糊控制器应用到实际的过程中，发现了一些问题，分析了这些问题产生的原因，并提出了改进方案，取得了良好的控制效果。

Chapter 9 BP – PID Controller Design and Experiment

第 9 章 神经网络 PID 控制器的设计实验

In the classical automatic control method, the design and implementation of the controller are mainly based on the root locus and frequency characteristics analysis. With the development of intelligent control methods, "Intelligent Control" as an course is given out in undergraduate teaching. As one of the main methods of intelligent control, neural network is used to design the controller very popular. So, in this chapter, the neural network intelligent controller design is studied for the experiment teaching.

在经典自动控制方法中，主要由根轨迹、频率特性等校正方法来实现控制器设计与实现。随着智能化方法的发展，"智能控制"作为一门课程也在本科教学中开设。智能控制的主要方法之一是运用神经网络法来设计控制器。与此相适应，本章主要研究神经网络智能化控制器的设计方法在实验教学中的应用。

9.1 Neural Network Control (NNC)

9.1 神经网络控制

9.1.1 The Methodology of NNC

9.1.1 神经网络控制理论

Artificial neural network is composed of a large number of artificial neurons. Usually, the information processing system composed by neural network is called artificial neural network system. With the development of industry, the plant object becomes more and more complex, and there is much more requirement for the control system, especially for some nonlinear system which is difficult to control, ordinary linear control methods have been difficult to meet the requirements. The artificial neural network has the ability to approach any complex nonlinear system, and it can adapt to the dynamic characteristics of

神经网络由大量的人工神经元相互连接而成，通常把由人工神经元组成的信息处理系统称为人工神经网络系统。随着时代的发展，被控对象的复杂程度越来越高，人们对控制系统的要求也越来越高，特别是针对一些难以控制的非线性系统，普通的线性控制方法已经难以满足要求。而人工神经网络有着处理任意复杂非线性系统的能力，并且可以适应不确定系统的动态特性。所以，运用人工神经网络设计控制器就显出其优越性。

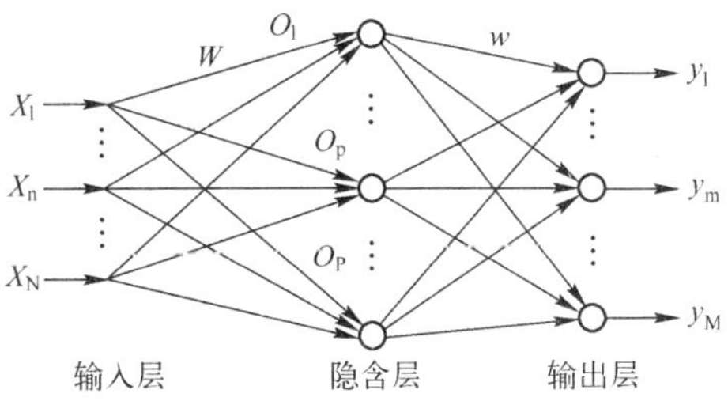

Figure 9.1 The Structure of BP Neural Network

图 9.1 BP 神经网络结构图

the uncertain system. Therefore, the use of artificial neural network controller design shows its superiority

This chapter discusses how to use BP neural network to design controller. BP neural network means the "error back propagation neural network", as shown is Figure 9.1, it is a kind of one – way propagation and multilayer forward network, the gradient descent method is used to minimize the mean square error between the desired output and the actual output. The learning process of BP network includes both forward and backward propagation. In the process of forward propagation, the input value passes through the hidden layer to the output layer, and the neuron state of each layer only affects the neurons in the next layer. If the final output of the output layer does not match the expected output, that means the actual output and the expected output hare error, the network will be back propagation, and return the error signal along the path of forward propagation, the neural network is updated by modifying the weights corresponding to each node of each layer, and then into the forward propagation, so cycle calculation, until the error of actual output and the expected output meets the requirement.

本章讨论选用 BP 神经网络来设计控制器。BP 神经网络的全称是误差反向传播神经网络（Back Propagation），如图 9.1 所示，它是一种单向传播的多层前向网络，采用梯度下降的最优化方法，使得网络的期望输出值与实际输出值的误差均方值达到最小。BP 网络的学习过程包含正向传播和反向传播。在正向传播过程中，输入值经过隐含层的处理传向输出层，并且每层的神经元状态只影响下一层的神经元。如果输出层得到的最终输出与期望输出不符，即实际输出与期望输出存在误差时，网络便会进行反向传播，把误差信号沿着正向传播的路径返回，通过修改各层各个节点对应的权值来更新神经网络，然后进入正向传播，如此反复，直到实际输出与期望输出的误差达到要求为止。

Thus, BP neural network is a supervised learning process. The structure of neural network is shown in figure 9. 1, there are M neurons existed in input layer, Q neurons in hidden layer, and L neurons in output layer when we see figure 9. 1 from left to right. The input of the j node of the hidden layer can be known from the figure 9. 1:

$$net_j = \sum_{i=1}^{M} \omega_{ij} O_i \tag{9.1}$$

The outputof the j node of the hidden layer is as following:

$$O_j = f(net_j) \tag{9.2}$$

f (·) is exciting function:

$$f(net_j) = \frac{1}{1 + e^{-\frac{net_j - \theta_j}{\theta_0}}} \tag{9.3}$$

When θ_j is positive, it makes the right shift of exciting function $f(net_j)$; and the θ_0 will adjust the shape of $f(net_j)$, the smaller the θ_0, the steeper the function $f(net_j)$.

Same as the input of the hidden layer, the input of the k node of the output layer is as following:

$$net_k = \sum_{i=1}^{Q} \omega_{jk} O_j \tag{9.4}$$

The output of network is as following:

$$O_k = f(net_k) \tag{9.5}$$

Assume the quadratic error function is as the following:

$$E = \frac{1}{2} \sum_{k=1}^{L} (d_k - O_k) \tag{9.6}$$

The weights are adjusted according to the inverse direction of the gradient of the E function, which makes the actual output approximating to the desired output. The correction formula of the output layer weight coefficient is as following:

由此可见，BP 神经网络是一个有导师的学习过程。神经网络结构如图 9. 1 所示，由左至右分别为输入层 M 个神经元，隐含层 Q 个神经元和输出层 L 个神经元。由图可以知道隐含层的第 j 个节点的输入为：

$$net_j = \sum_{i=1}^{M} \omega_{ij} O_i \tag{9.1}$$

而其输出为

$$O_j = \mathrm{f}(net_j) \tag{9.2}$$

f(·)为激发函数

$$\mathrm{f}(net_j) = \frac{1}{1 + e^{-\frac{net_j - \theta_j}{\theta_0}}} \tag{9.3}$$

其中的 θ_j 为正值时，其作用是使得激发函数 $f(net_j)$ 整体右移；θ_0 是起到调节 $f(net_j)$ 函数的形状，θ_0 越小 $f(net_j)$ 函数越陡。

与隐含层输出同理，输出层第 k 个节点的输入为：

$$net_k = \sum_{i=1}^{Q} \omega_{jk} O_j \tag{9.4}$$

网络的输出为：

$$O_k = f(net_k) \tag{9.5}$$

设样本的输入输出二次型误差函数定义为：

$$E = \frac{1}{2} \sum_{k=1}^{L} (d_k - O_k) \tag{9.6}$$

权系数按 E 函数的梯度的反方向进行调整即可使得实际输出向期望输出逼近。输出层权系数的修正公式为：

$$\Delta\omega_{jk} = -\eta \frac{\partial E}{\partial net_k}\frac{\partial net_k}{\partial \omega_{jk}} \tag{9.7}$$

Define the error back propagation is:

$$\delta_k = -\frac{\partial E}{\partial net_k} = -\frac{\partial E}{\partial O_k}\frac{\partial O_k}{\partial net_k} \tag{9.8}$$

where

$$\frac{\partial E}{\partial O_k} = -(d_k - O_k) \tag{9.9}$$

$$\frac{\partial O_k}{\partial net_k} = \frac{\partial[f(net_k)]}{\partial net_k} = f'(net_k) \tag{9.10}$$

So we have equation (9.11):

$$\begin{aligned}\delta_k &= (d_k - O_k)f'(net_k)\\ &= O_k(1-O_k)(d_k-O_k)\end{aligned} \tag{9.11}$$

and

$$\frac{\partial net_k}{\partial \omega_{jk}} = \frac{\partial}{\partial \omega_{jk}}\left(\sum_{j=1}^{Q}\omega_{jk}O_j\right) \tag{9.12}$$

according to the above equations, we can get the correction formula of the neuron weights in output layer:

$$\begin{aligned}\Delta\omega_{jk} &= \eta(d_k - O_k)f'(net_k)O_j\\ &= \eta\delta_k O_j\end{aligned} \tag{9.13}$$

The adjustment formula of weight coefficient in hidden layer is as following:

$$\begin{aligned}\Delta\omega_{ij} &= -\eta \frac{\partial E}{\partial net_k}\frac{\partial net_k}{\partial \omega_{jk}}\\ &= \eta\left(-\frac{\partial E}{\partial O_j}\right)f'(net_j)O_i = \eta\delta_j O_i\end{aligned} \tag{9.14}$$

the partial derivative cannot be calculated directly, the other indirect terms is necessary, that is:

$$-\frac{\partial E}{\partial O_j} = -\sum_{k=1}^{L}\frac{\partial E}{\partial net_k}\frac{\partial net_k}{\partial O_j} = \sum_{k=1}^{L}\delta_k\omega_{jk} \tag{9.15}$$

so, the output layer weight adjustment equation is:

$$\Delta\omega_{jk} = -\eta \frac{\partial E}{\partial net_k}\frac{\partial net_k}{\partial \omega_{jk}} \tag{9.7}$$

将误差反传信号定义为：

$$\delta_k = -\frac{\partial E}{\partial net_k} = \frac{\partial E}{\partial O_k}\frac{\partial O_k}{\partial net_k} \tag{9.8}$$

其中

$$\frac{\partial E}{\partial O_k} = -(d_k - O_k) \tag{9.9}$$

$$\frac{\partial O_k}{\partial net_k} = \frac{\partial}{\partial net_k}f(net_k) = f'(net_k) \tag{9.10}$$

因此我们可以得到式 (9.11)：

$$\begin{aligned}\delta_k &= (d_k - O_k)f'(net_k)\\ &= O_k(1-O_k)(d_k-O_k)\end{aligned} \tag{9.11}$$

又有，

$$\frac{\partial net_k}{\partial \omega_{jk}} = \frac{\partial}{\partial \omega_{jk}}\left(\sum_{j=1}^{Q}\omega_{jk}O_j\right) \tag{9.12}$$

由上几式可以得到输出层的神经元权系数的修正公式为：

$$\begin{aligned}\Delta\omega_{jk} &= \eta(d_k - O_k)f'(net_k)O_j\\ &= \eta\delta_k O_j\end{aligned} \tag{9.13}$$

隐含层权系数的调整公式为：

$$\begin{aligned}\Delta\omega_{ij} &= -\eta \frac{\partial E}{\partial net_k}\frac{\partial net_k}{\partial \omega_{jk}}\\ &= \eta\left(-\frac{\partial E}{\partial O_j}\right)f'(net_j)O_i = \eta\delta_j O_i\end{aligned} \tag{9.14}$$

其中的偏导项并不能直接进行计算，而需要其他的间接量，即：

$$-\frac{\partial E}{\partial O_j} = -\sum_{k=1}^{L}\frac{\partial E}{\partial net_k}\frac{\partial net_k}{\partial O_j} = \sum_{k=1}^{L}\delta_k\omega_{jk} \tag{9.15}$$

最终得到输出层的权值调整式为：

$$\Delta\omega_{jk} = \eta O_k(d_k - O_k)O_j \quad (9.16)$$

So, the hidden layer weight adjustment equation is:

$$\Delta\omega_{ij} = \eta(1 - O_j)\left(\sum_{k=1}^{L}\delta_k\omega_{jk}\right)O_i \quad (9.17)$$

Through the above equations, we can get the weights of the network connection as following:

$$\omega_{ij}(t+1) = \omega_{ij}(t) + \eta\delta_j O_i \quad (9.18)$$

$$\omega_{jk}(t+1) = \omega_{jk}(t) + \eta\delta_k O_j \quad (9.19)$$

9.1.2 The Design of BP - PID Controller

The PID controller has simple structure, clear physical meaning of each parameter, the model dependence is small, and is easy to realize in practice, the conventional PID controller is, even in the new control theory emerging today, the most commonly used methods in industrial process control. However, the main limitation of PID controller is its dependence on the controlled object. Due to the standard for controller performance are higher and higher, object is more and more complex (long time delay, time - varying, nonlinear complex system), only using the conventional PID controller is difficult to meet the system requirements, the neural network BP - PID controller is a useful control strategy for these problems.

$$\Delta\omega_{jk} = \eta O_k(d_k - O_k)O_j \quad (9.16)$$

隐含层的权值调整时为:

$$\Delta\omega_{ij} = \eta(1 - O_j)\left(\sum_{k=1}^{L}\delta_k\omega_{jk}\right)O_i \quad (9.17)$$

通过上述公式可以得到网络连接权值的调整式为:

$$\omega_{ij}(t+1) = \omega_{ij}(t) + \eta\Delta\omega_{ij} \quad (9.18)$$

$$\omega_{jk}(t+1) = \omega_{jk}(t) + \eta\Delta\omega_{jk} \quad (9.19)$$

9.1.2 BP - PID 控制器设计

由于 PID 控制器具有结构简单，各参数物理意义明确，对模型依赖程度小和工程上易于实现等优点，使得常规 PID 控制器，即使在新控制理论不断涌现的今天，仍是工业过程控制中最常用的方法。但是，PID 控制器主要的局限在于它对被控制对象的依赖性。由于对控制器品质要求越来越高，控制对象越来越复杂（大滞后、时变、非线性复杂系统），单纯采用常规的 PID 控制器很难满足系统要求，神经网络 BP - PID 控制是针对上述问题而提出的一种控制策略。

BP - PID is composed of PID control law into the neural network, and realizes the essential combination of neural network and PID control law. The hidden layer of BP - PID consists of 3 parts, which are proportional, integral and differential. It is a kind of dynamic forward network, which is more suitable for the design of control system. The number of neurons in each layer of the BP - PID, the connection mode and the connection weights are determined according to the basic principle of PID control law and the existing experience, to ensure the system stability and fast convergence. BP - PID not only has the advantages of traditional PID controller, but also has the parallel structure of neural network and the function of learning and memory, and ability to approximate arbitrary functions of multilayer networks

BP - PID 是将 PID 控制规律融进神经网络之中构成的，实现了神经网络和 PID 控制规律的结合。BP - PID 的隐含层由比例、积分、微分 3 个单元组成，是一种动态前向网络，更适于控制系统的设计。BP - PID 的各层神经元个数、连接方式、连接权值是按 PID 控制规律的基本原则和已有的经验确定的，保证了系统稳定和快速收敛。BP - PID 既具有传统 PID 控制器的优点，又具有神经网络的并行结构和学习记忆功能，及多层网络逼近任意函数的能力。

Based on the structure and algorithm of BP - PID controller, the convergence and stability of BP - PID control system are analyzed and proved, finally, simulation examples and conclusions are given.

下面将在给出 BP - PID 控制器的结构和算法的基础上，分析和证明 BP - PID 控制系统的收敛性和稳定性，最后给出仿真实例和结论。

9.2 BP-PID Controller Structure and Algorithm

9.2 BP - PID 控制器结构和算法

9.2.1 BP - PID Controller Structure

9.2.1 BP - PID 控制器结构

As shown in Figure 9.2, the BP - PID controller is a 3 - layer network, including input layer, hidden layer and output layer, structure is 2 - 3 - 1. The input layer of the network consists of 2 neurons, which are the input value of the control system r and the output value of the controlled object

如图 9.2 所示，BP - PID 控制器是 3 层网络，包括输入层、隐含层、输出层，其结构为 2—3—1。网络的输入层由 2 个神经元构成，分别输入控制系统的给定值 r 和被控对象输出值 y；隐含层有 3 个神经元，各神经元的输出函数互不相同，分别对应于比例（P）、积分（I）、微分

y. There are 3 neurons in the hidden layer, and the output function of each neuron is different from each other, which corresponds to the ratio (P), integral (I), differential (D) of the 3 units. The output layer of the network has 1 variable, which is integrated with BP – PID control law. The forward calculation of the network achieve BP – PID control law, the network reverse algorithm realizes adaptive adjustment of BP – PID parameters.

(D) 3 个单元；网络的输出层只有 1 个变量 u，完成 BP – PID 控制规律的综合。网络的前向计算实现 BP – PID 控制规律，网络的反向算法实现 BP – PID 参数的自适应调整。

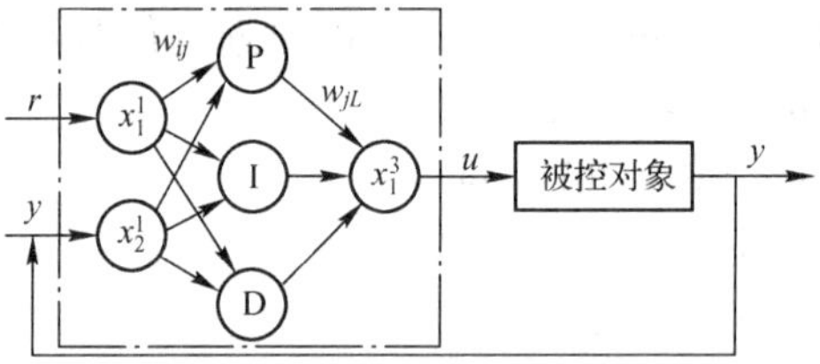

Figure 9.2　Diagram of BP – PID Neural Network Controller Structure

图 9.2　BP – PID 神经网络控制器结构图

9.2.2　BP – PID Control Algorithm

(1) BP – PID calculate method

The input layer and output layer of BP – PID are similar to those of general multilayer feedforward neural network. According to the given value of the control system and the output value of the controlled object plant, there is

$$x_1^1 = r(k)$$

$$x_2^1 = y(k)$$

The input function of each neuron in the hidden layer is:

$$I_j^2 = \sum_{i=1}^{2} w_{i,j} x_i^1(k) \quad j = 1,2,3$$

9.2.2　BP – PID 的控制算法

(1) BP – PID 的计算方法

BP – PID 的输入层和输出层与一般多层前向神经网络类似。根据控制系统给定值和被控对象输出值，有：

$$x_1^1 = r(k)$$

$$x_2^1 = y(k)$$

隐含层各神经元的输入函数为：

$$I_j^2 = \sum_{i=1}^{2} w_{i,j} x_i^1(k) \quad j = 1,2,3$$

The output function of the hidden layer is different, which is divided into the proportion, integral, differential of three units. In the actual control system, the output energy is limited as finite value, so the BP – PID's output is processecl with an upper and lower amplitude limitation. In order to prevent supersaturating, the integral unit is also limited for its amplitude.

The state of a proportional unit is:

$$x_1^2 = I_1^2(k) \tag{9.20}$$

The state of integral unit is:

$$x_2^2 = x_2^2(k-1) + I_2^2(k) \tag{9.21}$$

The state of differential unit is:

$$x_3^2 = I_3^2(k) - I_3^2(k-1) \tag{9.22}$$

Output of neuron output layer is:

$$x_1^3 = \sum_{j=1}^{3} w_{j,1} x_j^2(k) \tag{9.23}$$

Where w_{ij} are the weights from the input layer to the hidden layer; w_{j1} are the weights between the hidden layer and the output layer; k is sample interval; $i=1,2$ is the BP – PID serial number of input layer neurons; $j=1,2$ is the BP – PID serial number of hidden layer neurons.

(2) Back propagation algorithm of BP – PID control system

The back propagation algorithm of BP – PID control system completes the modification of network weights, realize the learning and memory functions in neural network. The goal of learning is to make $J = E_p = \frac{1}{m}\sum_{k=1}^{m}[r(k) - y(k)]^2$ reaching the minimum. Where r is the given value of the system, y is output value for the system, m is sum of sampling points.

隐含层的输出函数各不相同，分为比例元、积分元、微分元。考虑到在实际控制系统中，控制器输出能量为有限值，因此此处对 BP – PID 输出进行了上下限幅处理。为防止过饱和，对积分单元也进行了限幅处理。

比例元的状态为：

$$x_1^2 = I_1^2(k) \tag{9.20}$$

积分元的状态为：

$$x_2^2 = x_2^2(k-1) + I_2^2(k) \tag{9.21}$$

微分元的状态为：

$$x_3^2 = I_3^2(k) - I_3^2(k-1) \tag{9.22}$$

输出层神经元的输出为：

$$x_1^3 = \sum_{j=1}^{3} w_{j,1} x_j^2(k) \tag{9.23}$$

其中：w_{ij} 为输入层至隐含层的连接权值；w_{j1} 为隐含层至输出层的连接权值；k 为采样时刻；$i=1$，2 为 BP – PID 中输入层神经元序号；$j=1$，2，3 为 BP – PID 中隐含层神经元序号。

(2) BP – PID 控制系统的反传算法

BP – PID 控制系统的反传算法完成网络权值的修改、神经网络的学习和记忆功能。学习的目标是使 $J = E_p = \frac{1}{m}\sum_{k=1}^{m}[r(k) - y(k)]^2$ 为最小，式中 r 为系统给定值，y 为系统输出值，m 为每批采样点数。

According to the gradient method to adjust the BP - PID weight value, set the learning step, after the n step training, the hidden layer to the output layer weights are:

$$w_{j1}(n_0+1)=w_{j1}(n_0)-\eta_{j1}\frac{\partial J}{\partial w_{j1}} \tag{9.24}$$

The change of weight between hidden layer to the output layer is

$$\begin{aligned}\frac{\partial J}{\partial w_{j1}} &= \frac{\partial J}{\partial y}\cdot\frac{\partial y}{\partial x_1^3}\cdot\frac{\partial x_1^3}{\partial w_{j1}} \\ &= -\frac{2}{m}\sum_{k=1}^{m}[r(k)-y(k)] \\ &\quad \frac{y(k+1)-y(k)}{x_1^3(k)-x_1^3(k-1)}x_j^2(k) \\ &= -\frac{1}{m}\sum_{k=1}^{m}\delta'_{j1}(k)x_j^2(k)\end{aligned} \tag{9.25}$$

where

$$\delta'_{j1}=2[r(k)-y(k)]\cdot\frac{y(k+1)-y(k)}{x_1^3(k)-x_1^3(k-1)}$$

And $x_1^3(k)$ is the output vector of neurons output layer.

The $\frac{y(k+1)-y(k)}{x_1^3(k)-x_1^3(k-1)}$ is used to replace the $\frac{\partial y}{\partial x_1^3}$ in order to solve the question caused by unknown object parameters. It can be seen that this term is only one multiplication factor, and the positive and negative of the term determines the direction of convergence. The size of this term value only determines the rate of convergence.

The weight of the input layer to the hidden layer is:

$$w_{ij}(n_0+1)=w_{ij}(n_0)-\eta_{j1}\frac{\partial J}{\partial w_{j1}} \tag{9.26}$$

按梯度法调节 BP - PID 权值，设定学习步长，经过 n 步训练后，隐含层至输出层的权重值为：

$$w_{j1}(n_0+1)=w_{j1}(n_0)-\eta_{j1}\frac{\partial J}{\partial w_{j1}} \tag{9.24}$$

隐含层对输出层的权值变化为：

$$\begin{aligned}\frac{\partial J}{\partial w_{j1}} &= \frac{\partial J}{\partial y}\cdot\frac{\partial y}{\partial x_1^3}\cdot\frac{\partial x_1^3}{\partial w_{j1}} \\ &= -\frac{2}{m}\sum_{k=1}^{m}[r(k)-y(k)] \\ &\quad \frac{y(k+1)-y(k)}{x_1^3(k)-x_1^3(k-1)}x_j^2(k) \\ &= -\frac{1}{m}\sum_{k=1}^{m}\delta'_{j1}(k)x_j^2(k)\end{aligned} \tag{9.25}$$

其中：

$$\delta'_{j1}(k)=2[r(k)-y(k)]\cdot\frac{y(k+1)-y(k)}{x_1^3(k)-x_1^3(k-1)}$$

式中 $x_1^3(k)$ 为输出层神经元的输出值。

在式中用 $\frac{y(k+1)-y(k)}{x_1^3(k)-x_1^3(k-1)}$ 项取代 $\frac{\partial y}{\partial x_1^3}$，以解决对象参数未知造成的困难。由此式可知，该项仅为其中的一个相乘因子，该项的正负决定收敛方向，其中大小只决定收敛的速度。

输入层对隐含层的权重值为：

$$w_{ij}(n_0+1)=w_{ij}(n_0)-\eta_{j1}\frac{\partial J}{\partial w_{j1}} \tag{9.26}$$

Similarly, it is known that the weight change of the input layer to the hidden layer is

$$\frac{\partial J}{\partial w_{ij}} = \frac{\partial J}{\partial y} \cdot \frac{\partial y}{\partial x_1^3} \cdot \frac{\partial x_1^3}{\partial x_j^2} \cdot \frac{\partial x_j^2}{\partial I_j^2} \cdot \frac{\partial I_j^2}{\partial w_{ij}} = -\frac{1}{m}\sum_{k=1}^{m} \delta_{j1} x_1^1(k) \tag{9.27}$$

where

$$\delta_{j1} = \delta'_{j1} \cdot w_{j1} \frac{x_j^2(k+1) - x_j^2(k)}{I_j^2(k) - I_j^2(k-1)} \tag{9.28}$$

(3) The initial weight value of the BP – PID and the input weight of the integral unit

Selecting the appropriate initial weights of the network can accelerate the learning and convergence speed, and the initial value of the weight of BP – PID is selected according to the characteristics of PID control law.

The initial value of the input layer to the hidden layer satisfies the requirement for mapping of $(r, y) \to e$:

$r \to$ hidden layer, $w_{1j}(0) = +1$

$y \to$ hidden layer, $w_{2j}(0) = -1$

Set the initial weight of hidden layer to the output layer to a small positive without prior experience of the situation, select $w_{j1} = +0.01$. At the same time, in order to ensure the system adjustment without static error, input layer to the hidden layer weights integral unit values $(r, y) \to e$ should meet to no distortion. Therefore, the input layer to the hidden layer integral unit of the weight value, the absolute value w_{11}, w_{12} in the entire process are the same, the marks are opposite.

同理可知输入层至隐含层的权植变化量为：

$$\frac{\partial J}{\partial w_{ij}} = \frac{\partial J}{\partial y} \cdot \frac{\partial y}{\partial x_1^3} \cdot \frac{\partial x_1^3}{\partial x_j^2} \cdot \frac{\partial x_j^2}{\partial I_j^2} \cdot \frac{\partial I_j^2}{\partial w_{ij}} = -\frac{1}{m}\sum_{k=1}^{m} \delta_{j1} x_1^1(k) \tag{9.27}$$

其中：

$$\delta_{j1} = \delta'_{j1} \cdot w_{j1} \frac{x_j^2(k+1) - x_j^2(k)}{I_j^2(k) - I_j^2(k-1)} \tag{9.28}$$

(3) BP – PID 权重初值的设定及积分单元输入权值的处理

选择合适的网络权重初值，可以加快学习和收敛速度，BP – PID 的权重初值按 PID 控制规律的特点选取。

输入层至隐含层权重初值满足映射 $(r, y) \to e$ 的要求：

$r \to$ 隐含层 $w_{1j}(0) = +1$

$y \to$ 隐含层 $w_{2j}(0) = -1$

隐含层至输出层的权重初值在无先验经验的情况下设定为小正数，现取 $w_{j1} = +0.01$。同时，为了保证系统调节无静差，输入层至隐含层积分单元的权重值应满足 $(r, y) \to e$ 的无畸变性。所以，输入层至隐含层积分单元的权重值 w_{11}、w_{12} 的绝对值在整个过程中相等，符号相反。

The learning end condition:

Calculate the mean square error, the error equals the output of plant subtracting the system reference value in each T_p period m samples . If the mean square error is less than the allowable error, the learning stops.

学习结束的条件:

计算每个 T_p 时间内 m 个采样点的被控对象输出值和系统给定值的误差平方均值，当本次 T_p 时间内的误差平方均值小于设定允许误差时停止学习。

9.2.3 Convergence and Stability Analysis of BP – PID Controller

9.2.3 采用 BP – PID 控制器的收敛性和稳定性分析

(1) Convergence

The convergence of the BP – PID control system depends on the choice of the learning step. The following is the use of the Lyapunov stability principle to determine the range of learning steps to ensure the convergence of the system.

Determine the Lyapunov function

$$L(n_0) = \frac{1}{2}\bar{E}^2(n_0) \tag{9.29}$$

where

$$\bar{E}(n_0) = \sqrt{\frac{1}{m}\sum_{k=1}^{m}[r(k) - y(k)]^2} = \sqrt{J} \tag{9.30}$$

$r(k)$ is the reference input, and the $y(k)$ is the output of the system.

Where

$$\bar{E}(n_0+1) = \bar{E}(n_0) + \Delta\bar{E}(n_0) \tag{9.31}$$

the variable of error is:

$$\Delta\bar{E}(n_0) = \frac{\partial\bar{E}(n_0)}{\partial w}\cdot\Delta w \tag{9.32}$$

The variable of weight in network is

$$\Delta w = -\eta\frac{\partial J}{\partial w} = -\eta\frac{\partial}{\partial w}[\bar{E}^2(n_0)]$$
$$= -2\eta\bar{E}(n_0)\frac{\partial}{\partial w}[\bar{E}(n_0)] \tag{9.33}$$

(1) 收敛性

BP – PID 控制系统的收敛性取决于学习步长的选取，下面利用 Lyapunov 稳定性原理确定保证系统收敛的学习步长范围。

定义 Lyapunov 函数

$$L(n_0) = \frac{1}{2}\bar{E}^2(n_0) \tag{9.29}$$

其中:

$$\bar{E}(n_0) = \sqrt{\frac{1}{m}\sum_{k=1}^{m}[r(k) - y(k)]^2} = \sqrt{J} \tag{9.30}$$

式中，$r(k)$为系统被控变量给定值，$y(k)$为系统被控变量实际输出值。

其中:

$$\bar{E}(n_0+1) = \bar{E}(n_0) + \Delta\bar{E}(n_0) \tag{9.31}$$

误差变化量

$$\Delta\bar{E}(n_0) = \frac{\partial\bar{E}(n_0)}{\partial w}\cdot\Delta w \tag{9.32}$$

网络权值变化量

$$\Delta w = -\eta\frac{\partial J}{\partial w} = -\eta\frac{\partial}{\partial w}[\bar{E}^2(n_0)]$$
$$= -2\eta\bar{E}(n_0)\frac{\partial}{\partial w}[\bar{E}(n_0)] \tag{9.33}$$

Put the above formulae into equation (9.29), it can be obtained:

$$\Delta L(n_0) = \frac{1}{2}[\bar{E}^2(n_0+1) - \bar{E}^2(n_0)]$$
$$= \frac{1}{2}\{[\bar{E}(n_0) + \Delta\bar{E}(n_0)]^2 - \bar{E}^2(n_0)]$$
$$= 2[\bar{E}(n_0)]^2[-\eta + (\eta)^2] \tag{9.34}$$

In order to guarantee the convergence of the system, $\Delta L(n_0) < 0$ must be satisfied, that is:

$$\eta[\eta(\frac{\partial\bar{E}(n_0)}{\partial w})^2 - 1] < 0 \tag{9.35}$$

Solve the inequality, that is

$$0 < \eta < \frac{1}{(\partial\bar{E}(n_0)/\partial w)^2} \tag{9.36}$$

From formula (9.30), that is

$$\frac{\partial\bar{E}(n_0)}{\partial w} = \frac{\partial J}{2\sqrt{J}\,\partial w} \tag{9.37}$$

So, the convergence condition of BP – PID control system is:

$$0 < \eta < 1/\varepsilon^2$$

where

$$\varepsilon = -\frac{\partial J}{2\sqrt{J}\,\partial w}$$

Weight adjustment algorithm:

$$w(n_0+1) = w(n_0) - \eta\,\partial J/\partial w \tag{9.38}$$

The following two conclusions can be obtained from the above convergence conditions:

Conclusion 1 If the BP – PID control system is convergent, the learning step in the weight adjustment algorithm

$$w_{j1}(n_0+1) = w_{j1}(n_0) - \eta_{j1}\frac{\partial J}{\partial w_{j1}}$$

将以上几式代入式（9.29）中可得到：

$$\Delta L(n_0) = \frac{1}{2}[\bar{E}^2(n_0+1) - \bar{E}^2(n_0)]$$
$$= \frac{1}{2}\{[\bar{E}(n_0) + \Delta\bar{E}(n_0)]^2 - \bar{E}^2(n_0)]$$
$$= 2[\bar{E}(n_0)]^2[-\eta + (\eta)^2] \tag{9.34}$$

为了保证系统的收敛性，必须满足 $\Delta L(n_0) < 0$，即

$$\eta\left[\eta(\frac{\partial\bar{E}(n_0)}{\partial w})^2 - 1\right] < 0 \tag{9.35}$$

解不等式可得：

$$0 < \eta < \frac{1}{(\partial\bar{E}(n_0)/\partial w)^2} \tag{9.36}$$

由式（9.30），有

$$\frac{\partial\bar{E}(n_0)}{\partial w} = \frac{\partial J}{2\sqrt{J}\,\partial w} \tag{9.37}$$

即 BP – PID 控制系统收敛条件为：

$$0 < \eta < 1/\varepsilon^2$$

其中

$$\varepsilon = -\frac{\partial J}{2\sqrt{J}\,\partial w}$$

权值调整算法：

$$w(n_0+1) = w(n_0) - \eta\,\partial J/\partial w \tag{9.38}$$

由以上收敛条件可得到以下两个结论：

结论 1 如果 BP – PID 控制系统是收敛的，则其隐含层至输出层调整算法中，式

$$w_{j1}(n_0+1) = w_{j1}(n_0) - \eta_{j1}\frac{\partial J}{\partial w_{j1}}$$

from the hidden layer to the output layer will be satisfied of $0<\eta_{j1}<1/\varepsilon_{j1}^2$.

Where

$$\varepsilon_{j1}=-\frac{\sum_{k=1}^{m}\delta'_{j1}(k)x_j^2(k)}{2\sqrt{m\sum_{k=1}^{m}[r(k)-y(k)]^2}} \tag{9.39}$$

Conclusion 2 If the BP - PID control system is convergent, the weight adjustment algorithm from the input layer to the hidden layer is:

$$w_{ij}(n_0+1)=w_{ij}(n_0)-\eta\frac{\partial J}{\partial w_{ij}} \tag{9.40}$$

the weight of the learning step η_{ij} in formula (9.40) must satisfg $0<\eta_{ij}<1/\varepsilon_{ij}^2$, where

$$\varepsilon_{ij}=-\frac{\sum_{k=1}^{m}\delta'_{j1}(k)x_i^1(k)}{2\sqrt{m\sum_{k=1}^{m}[r(k)-y(k)]^2}} \tag{9.41}$$

Conclusion 1 and conclusion 2 give the range of the learning step in each layer of BP - PID, and the convergence of the system can be guaranteed as long as the step size is taken into account. In practical application, for ease of calculation, we can take a positive step which is small enough, then adjust it according to the above two conclusions.

(2) The stability of BP - PID control system

的学习步长必然满足：$0<\eta_{j1}<1/\varepsilon_{j1}^2$

其中：

$$\varepsilon_{j1}=-\frac{\sum_{k=1}^{m}\delta'_{j1}(k)x_j^2(k)}{2\sqrt{m\sum_{k=1}^{m}[r(k)-y(k)]^2}} \tag{9.39}$$

结论2 如果BP - PID控制系统是收敛的，则其输入层至隐含层的权值调整算法：

$$w_{ij}(n_0+1)=w_{ij}(n_0)-\eta\frac{\partial J}{\partial w_{ij}} \tag{9.40}$$

式（9.40）中的权值调整学习步长η_{ij}必须满足$0<\eta_{ij}<1/\varepsilon_{ij}^2$，

其中：

$$\varepsilon_{ij}=-\frac{\sum_{k=1}^{m}\delta'_{j1}(k)x_i^1(k)}{2\sqrt{m\sum_{k=1}^{m}[r(k)-y(k)]^2}} \tag{9.41}$$

结论1和结论2给出了BP - PID各层权值学习步长的范围，只要步长在此范围内取值，就可以保证系统的收敛性。在实际应用中，为了计算简便起见，可先取步长为一个足够小的正数，再根据以上两个结论进行适当的调整。

（2）BP - PID控制系统的稳定性

When BP – PID connected weight is selected as the initial value, if the control system is stable, select the appropriate weights of each layer to training, ensure the convergence of the control system, so that the BP – PID control system is stable.

BP – PID 的连接权值取初值时，如果控制系统是稳定的，再选取合适的各层权值学习，保证控制系统是收敛的，则 BP – PID 控制系统在学习后是稳定的。

(3) Condition of learning stop

Calculate the mean square of error between output of plant and input value of the system in each period of T_p, m sample data, the learning will be ended after the mean square error is less than the allowable error.

(3) 学习结束条件

计算每个 T_p 时间内 m 个采样点的对象输出和系统给定值的误差平方均值，当本次 T_p 时间内的误差平方均值小于设定允许误差时停止学习。

From the above analysis, we can see that BP – PID is formed by the integration of neural network and PID control law. BP – PID is a kind of multilayer feedforward neural network with dynamic and static characteristics. The convergence speed of the network is fast and the stability of the system is guaranteed. BP – PID controller has units of proportional, integral and differential, so the control system has fast response, small overshoot, and has no static error. Even more, the weight value of BP – PID has a clear physical meaning, and its parameter setting need not to know the parameters and mathematical model of the controlled object plant. So it should be used for different plants.

由以上分析可以看出，BP – PID 是将神经网络和 PID 控制律融为一体而形成的。BP – PID 是一种内含动、静特性的多层前向神经网络，网络收敛速度快且保证了系统的稳定性。BP – PID 中比例、积分、微分单元，使 BP – PID 控制系统响应快、超调小、无静差。并且 BP – PID 的权重值具有明确的物理意义，其参数设定无须知道被控对象的具体参数及数学模型，对不同的对象具有适应性。

9.3 BP – PID Controller Design and Simulation

Take a second order system as an example. The discrete model is as following:

$$x(k)=\begin{bmatrix}0 & 1\\ -0.6321211 & \end{bmatrix}x(k-1)+\begin{bmatrix}0\\1\end{bmatrix}u(k)$$

$$y(k)=\begin{bmatrix}0.464241\\0.367879\end{bmatrix}x(k)$$

9.3 BP – PID 控制器的算法设计与仿真

以一个二阶系统为例，其离散化模型为：

$$x(k)=\begin{bmatrix}0 & 1\\ -0.6321211 & \end{bmatrix}x(k-1)+\begin{bmatrix}0\\1\end{bmatrix}u(k)$$

$$y(k)=\begin{bmatrix}0.464241\\0.367879\end{bmatrix}x(k)$$

Under the condition of initial weights, the step input signal is 4. Learning step size is 0.05, sampling period is 0.01, each learning picks up 135 points. In the initial control condition, (the weights of input are 1, 1, 1; −1, −1, −1; the weights of output are 0.1, 0.1, 0.1), the output response of system will be shown at figure 9.3 ~9.6.

在初始权值条件下，输入为 4 的阶跃信号。学习步长为 0.05，采样周期 0.01，每次学习采 135 个点。初始控制条件下（输入权值选取 1，1，1；−1，−1，−1；输出权值选取 0.1，0.1，0.1），系统响应图像如图 9.3 ~ 图 9.6 所示。

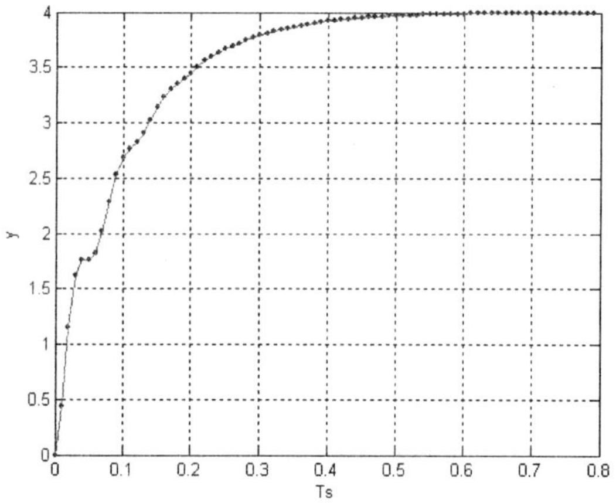

Figure 9.3 Simulation with Initial Condition

图 9.3 初始控制条件下的仿真图

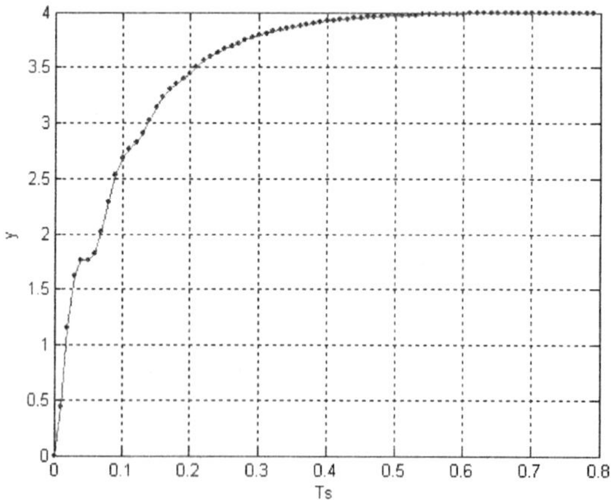

Figure 9.4 After 5 Times Learning Simulation

图 9.4 经过 5 次学习后的仿真图

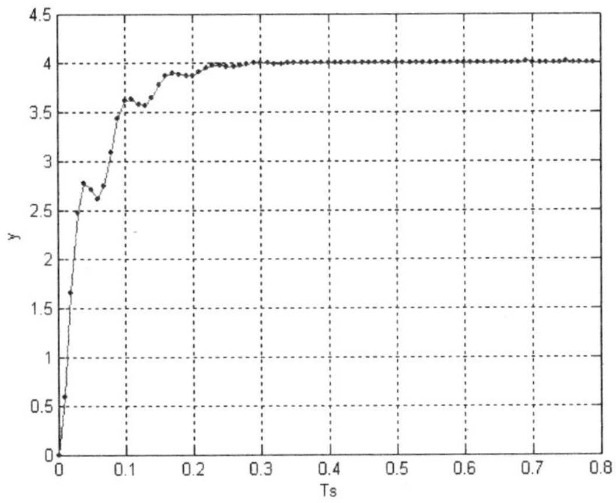

Figure 9. 5 after10 times learning simulation

图 9. 5 经过 10 次学习后的仿真图

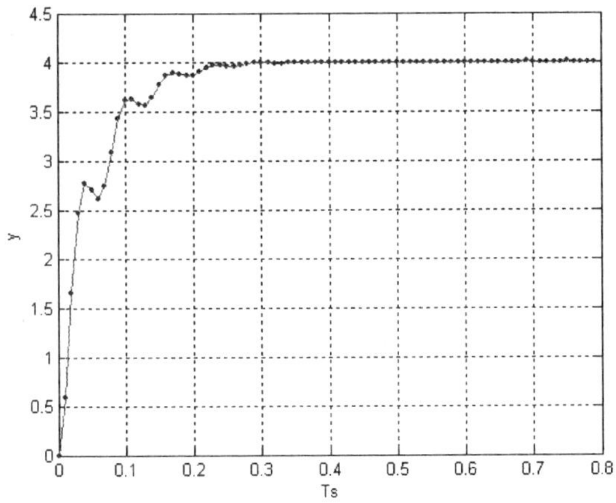

Figure 9. 6 after 15 times learning simulation

图 9. 6 经过 15 次学习后的仿真图

By comparing the 4 response curve, it is obviously that the output adjustment speed become fast. The more training times, the faster reach the steady state. But after many experiments, it can be found that the number of training group should not be too much. Large integration and proportion are not conducive to control stability.

通过对 4 组数据作比较，能够明显看出输出的调整速度加快，训练越多越快进入稳态过程。经过多次实验可发现训练组数不宜过多，过高的积分和比例环节都不利于控制稳定性。

9.4 Summary

This chapter introduces the basic theory of neural network control algorithms firstly, and then for the question of " PID controller can't deal with the large time delay, time varing, nonlinear system", the BP - PID neural network controller is in troduced how to design a new kind of controller. and the stability and convergence of the BP - PID controller are analyzed. Finally, the design and simulation results of the BP - PID neural network controller are given at the end of this chapter.

9.4 本章小结

本章首先介绍了神经网络控制算法的基本理论，然后针对常规 PID 控制器很难满足大滞后、时变、非线性复杂系统的控制要求等问题，介绍了如何用 BP - PID 神经网络控制器来设计一种新型控制器，分析了 BP - PID 控制器的稳定性和收敛性，最后给出了 BP - PID 神经网络控制器的设计与算法仿真的结果。

Chapter 10 Design of Controller Experiment with GA Method

第10章 基于遗传算法的控制器设计实验

Ball - beam system is a classical nonlinear unsteady plant, it is widely used for control algorithm examined in research and experiment. In university laboratory, the ball - beam system has become an experiment tool provided to students. Various algorithms such as PID, Fuzzy, Neural Network, and Genetic are used for Teaching and Experiment system by means of ball - beam system.

球杆系统是一种传统的非线性被控对象，广泛应用于研究与实验。大学的实验室也已经将球杆系统作为实验装置提供给学生，各类算法例如PID、模糊、神经网络、遗传算法等方法，都可以通过在球杆系统上进行实验获得算法验证。

Genetic algorithm (GA) is a global optimization technique. Marra M. A., et al. present a kind of GA controller with ball - beam system. The GA is used to optimize the performance of LQR controller. Liu Ke - ping, et al. try to combine GA with PID algorithm could improve the control performance.

遗传算法是一种全局优化技术。是由 Marra M. A. 等人提出的一种球杆系统上的遗传控制算法。用遗传算法来优化LQR控制器的性能指标。Liu Ke - ping 等人将遗传算法与PID算法结合，以促进控制性能。

Here a kind of GA controller is designed for ball - beam system. Because the ball - beam system is a classical nonlinear system, after the linearization, the state space mathematical model for the Ball - beam is established by using Lagrange equation. There are serious noises in the detection signal of ball - beam system, so the Kalman Filter is adopted to deal with the noise affection.

本书展示一种基于球杆系统的GA控制器设计，因为球杆系统是经典非线性系统，首先进行线性化，用拉格朗日方程进行数学建模。因为实验过程中有严重的噪声，故尝试用卡尔曼滤波方法对其进行滤除。

Since the ball - beam system is unsteady system originally, a suitable controller is necessary. PID controller is a popular control method, but we find it is difficult to reach a control state of "settling time, steady state error, and percent overshoot". So, we try to design an integral separation PID control algorithm based on genetic algorithms with a new kind of fitness function. The experimental data obtained shows the control effective comparing with classical PID controller.

因为球杆系统是先天不稳定的系统，因此需要设计合适的控制器。PID 控制器是一种广泛应用的控制器，但其控制指标很难达到"过渡过程时间，稳态误差，超调量"的要求。因此我们尝试设计一种基于遗传算法的积分分离的控制器，实验数据显示，该算法的控制效果比传统的 PID 控制器要好。

10.1 Modeling of Ball - beam System

10.1 球杆系统建模

10.1.1 The Dynamic Equation of Mechanism of Ball - beam System

10.1.1 球杆机械系统的动态方程

The structure of Ball - beam system is shown as Figure 10.1. Ball - beam is a popular experiment equipment for University Laboratory, the dynamic equation is obtained by Lagrange method as following:

$$\begin{cases} \left(\frac{J_b}{R^2}+m\right)\ddot{r}-m\dot{r}\alpha^2+mg\sin\alpha=0 \\ (J_l+mr^2)\ddot{\alpha}+2m\dot{r}r\dot{\alpha}+\left(mgr+\frac{1}{2}MgL\right) \\ \cos\alpha=\tau \end{cases} \quad (10.1)$$

Where

J_b is the moment inertia of ball; J_l is the moment inertia of beam; m is mass of ball ; M is mass of beam ; L is length of beam; R is radius of ball; d is radius of gear; g is acceleration of gravity; r is displacement of ball in beam; α is rotation angle of beam; θ is rotation angle of gear; τ is torque in beam. When the ball - beam system stays at balance state, there is $\alpha=0$. After linearization in balance state, there is an approximate linear formula as following:

球杆系统的结构图如图 10.1 所示。球杆系统是高校实验室的常用设备，其动态方程可以用拉格朗日法建立：

$$\begin{cases} \left(\frac{J_b}{R^2}+m\right)\ddot{r}-m\dot{r}\alpha^2+mg\sin\alpha=0 \\ (J_l+mr^2)\ddot{\alpha}+2m\dot{r}r\dot{\alpha}+\left(mgr+\frac{1}{2}MgL\right) \\ \cos\alpha=\tau \end{cases} \quad (10.1)$$

其中

J_b是球的惯性系数；J_l 是球杆的惯性系数；m 是球的质量；M 是杆的质量；L 是球杆长度；R 是球半径；d 是齿轮半径；g 是重力加速度；r 是球在球杆上的位置距离；α 是球杆旋转角度；θ 是齿轮转角；τ 是球杆转矩。当球杆系统处于状态平衡时，有 $\alpha=0$。在平衡点附近进行线性化，有下述近似方程成立：

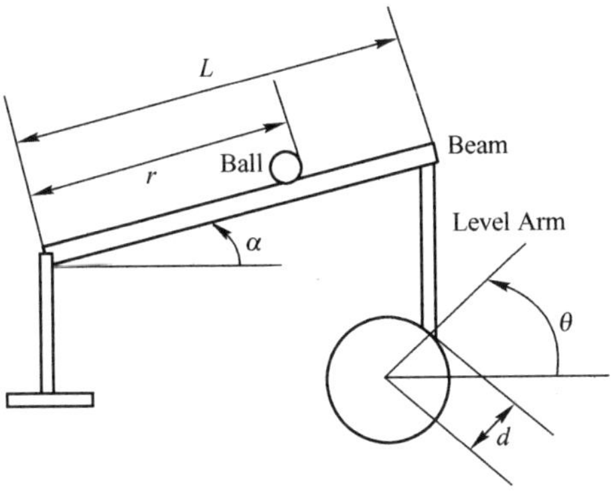

Figure 10. 1 Structure of Ball – beam System
图 10. 1 球杆系统结构

$$\left(\frac{J_b}{R^2}+m\right)\ddot{r}+mg\alpha=0 \quad (10.2)$$

$$\left(\frac{J_b}{R^2}+m\right)\ddot{r}+mg\alpha=0 \quad (10.2)$$

10. 1. 2 The Dynamic Equation of Beam Angle

The angle of gear rotation θ has the relation with the angle of beam α as following:

$$(L(1-\cos\alpha)-d(1-\cos\theta))^2+(L\sin\alpha+l-d\sin\theta)^2=l^2 \quad (10.3)$$

When the ball – beam stays at guide of beam with horizontal, there is an approximate equation as following:

$$\alpha=\frac{d}{L}\theta \quad (10.4)$$

10. 1. 2 球杆角度的动态方程

齿轮旋转角度 θ 与球杆角度 α 之间有以下的数学关系：

$$(L(1-\cos\alpha)-d(1-\cos\theta))^2+(L\sin\alpha+l-d\sin\theta)^2=l^2 \quad (10.3)$$

当球杆系统处于水平状态时，有下面的近似公式成立：

$$\alpha=\frac{d}{L}\theta \quad (10.4)$$

10. 1. 3 The Model of Ball – beam System

The model of ball – beam system is shown as Figure 10. 2.

10. 1. 3 球杆系统的数学模型

球杆系统的模型如图 10. 2 所示：

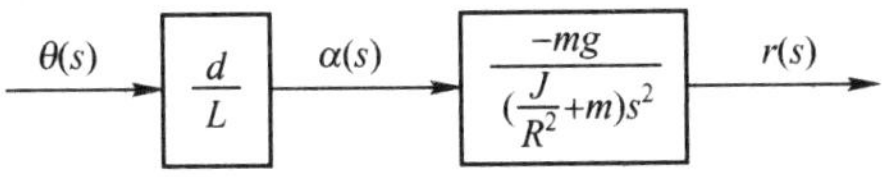

Figure 10. 2 Model of Ball – beam System
图 10. 2 球杆系统模型

The Model can be written as:

$$\frac{r(s)}{\theta(s)}=\frac{mgd}{L\left(\frac{J_b}{R^2}+m\right)}\frac{1}{s^2} \quad (10.5)$$

The parameters of formula (10.5) are known, and rewrite formula (10.5) as following:

$$\frac{r(s)}{\theta(s)}=\frac{0.7}{s^2} \quad (10.6)$$

球杆系统模型可以用下式表示：

$$\frac{r(s)}{\theta(s)}=\frac{mgd}{L\left(\frac{J_b}{R^2}+m\right)}\frac{1}{s^2} \quad (10.5)$$

式（10.5）中的参数为已知量，将其化简如下式所示：

$$\frac{r(s)}{\theta(s)}=\frac{0.7}{s^2} \quad (10.6)$$

10.2 Kalman Filter Design

According to Formula (10.5), there is

$$\ddot{r}=\frac{mg}{\left[m+\frac{J}{R^2}\right]}\alpha=7\alpha \quad (10.7)$$

assuming:

$$x_1=r$$

$$x_2=\dot{r}$$

we have the state equation:

$$\begin{cases}\dot{x}_1=x_2\\ \dot{x}_2=0.175\theta\end{cases} \quad (10.8)$$

Rearrange (10.8), we have equation (10.9) as following

$$\begin{bmatrix}\dot{x}_1\\ \dot{x}_2\end{bmatrix}=\begin{bmatrix}0 & 1\\ 0 & 0\end{bmatrix}\begin{bmatrix}x_1\\ x_2\end{bmatrix}+\begin{bmatrix}0\\ 0.175\end{bmatrix}\theta \quad (10.9)$$

After discretization (10.9), the new state and observe equation is shown as formula (10.10)

$$\begin{bmatrix}x_1(k)\\ x_2(k)\end{bmatrix}=\begin{bmatrix}1 & T\\ 0 & 1\end{bmatrix}\begin{bmatrix}x_1(k-1)\\ x_2(k-1)\end{bmatrix}+\begin{bmatrix}0.0875T^2\\ 0.175T\end{bmatrix}$$

$$\theta(k-1)r(k)=\begin{bmatrix}1 & 0\end{bmatrix}\begin{bmatrix}x_1(k)\\ x_2(k)\end{bmatrix}$$

(10.10)

10.2 卡尔曼滤波器设计

根据式（10.5），可得下列公式：

$$\ddot{r}=\frac{mg}{\left[m+\frac{J}{R^2}\right]}\alpha=7\alpha \quad (10.7)$$

假定：

$$x_1=r$$

$$x_2=\dot{r}$$

我们有下面状态方程：

$$\begin{cases}\dot{x}_1=x_2\\ \dot{x}_2=0.175\theta\end{cases} \quad (10.8)$$

简化式（10.8），得到式（10.9）：

$$\begin{bmatrix}\dot{x}_1\\ \dot{x}_2\end{bmatrix}=\begin{bmatrix}0 & 1\\ 0 & 0\end{bmatrix}\begin{bmatrix}x_1\\ x_2\end{bmatrix}+\begin{bmatrix}0\\ 0.175\end{bmatrix}\theta \quad (10.9)$$

将式（10.9）离散化，有式（10.10）成立：

$$\begin{bmatrix}x_1(k)\\ x_2(k)\end{bmatrix}=\begin{bmatrix}1 & T\\ 0 & 1\end{bmatrix}\begin{bmatrix}x_1(k-1)\\ x_2(k-1)\end{bmatrix}+\begin{bmatrix}0.0875T^2\\ 0.175T\end{bmatrix}$$

$$\theta(k-1)r(k)=\begin{bmatrix}1 & 0\end{bmatrix}\begin{bmatrix}x_1(k)\\ x_2(k)\end{bmatrix}$$

(10.10)

For Kalman filter, formula (10.10) will be written as formula (10.11):

$$\begin{bmatrix} x_1(k) \\ x_2(k) \end{bmatrix} = \begin{bmatrix} 1 & T \\ 0 & 1 \end{bmatrix} \begin{bmatrix} x_1(k-1) \\ x_2(k-1) \end{bmatrix} + \begin{bmatrix} 0.0875T^2 \\ 0.175T \end{bmatrix} [W_{k-1} + \theta(k-1)]$$

$$Z(k) = [1 \quad 0] \begin{bmatrix} x_1(k) \\ x_2(k) \end{bmatrix} + V_k \tag{10.11}$$

Where$Z(k)$ is observe vector, W_{k-1} is system noise, and the V_k is measurement noise. Formula (10.11) is a linear equation, it will be used for Kalman Filter directly.

The position of ball after Kalman filter is shown as Figure 10.3, and the signal noise reduced by Kalman filter is shown as Figure 10.4.

对于卡尔曼滤波器，式（10.10）可以写成式（10.11）的形式：

$$\begin{bmatrix} x_1(k) \\ x_2(k) \end{bmatrix} = \begin{bmatrix} 1 & T \\ 0 & 1 \end{bmatrix} \begin{bmatrix} x_1(k-1) \\ x_2(k-1) \end{bmatrix} + \begin{bmatrix} 0.0875T^2 \\ 0.175T \end{bmatrix} [W_{k-1} + \theta(k-1)]$$

$$Z(k) = [1 \quad 0] \begin{bmatrix} x_1(k) \\ x_2(k) \end{bmatrix} + V_k \tag{10.11}$$

其中 $Z(k)$ 为观测向量，W_{k-1} 是系统噪声，V_k 是观测噪声。式（10.11）是线性方程，可直接用于卡尔曼滤波。

经过卡尔曼滤波后的位置动态曲线图如图 10.3 所示。滤掉的噪声信号图如图 10.4 所示。

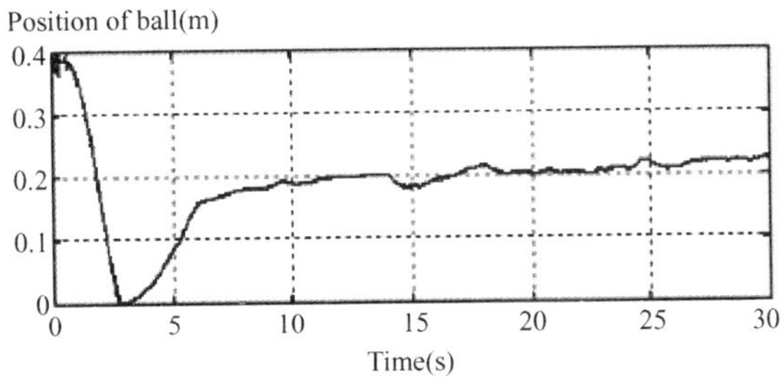

Figure 10.3 The Position of ball Dynamic Curve after Kalman Filter

图 10.3 经过卡尔曼滤波后的球的位置动态曲线

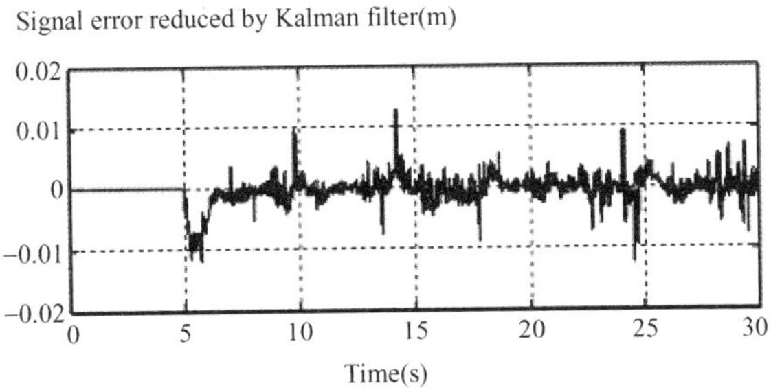

Figure 10.4 The Noise Deal with by Kalman Filter

图 10.4 经过卡尔曼滤波器滤除的噪声曲线

10.3 Controller Design

After the Kalman Filtering, we try to design a GA + PID controller. Assuming J is fitness function, determine J as following:

$$J = \int_0^\infty (w_1 |e(t)| + w_2 \sup(t) + w_3 u^2(t)) \mathrm{d}t + w_4 t_u + w_5 t_s \qquad (10.11)$$

Where $e(t)$ is system noise, $u(t)$ is output of controller, $\sup(t)$ is percent overshoot, t_u is rising time, t_s is settling time, w_1, w_2, w_3, w_4, w_5 are weights of system. According to real ball - beam system, they are selected as following:

$w_1 = 100$, $w_2 = 1000$, $w_3 = 0.1$,

$w_4 = 200$, $w_5 = 100$.

Adopt binary - coded, the initial population size is $M = 20$, genetic generation number $G = 70$, Roulette algorithm selection probability, and cross probability is $P_c = 0.7$, mutation probability is:

$P_m = 0.03 - [1:1:\mathrm{Size}] * (0.015)/M$.

Consider the practice ball - beam system, the parameters of $K_p, K_{i1}, K_{i2}, K_d, e_1, e_2$ will be as the vectors that are optimized. Determine the parameters as following:

1) when $|e(t)| < = e_1$, should select PID controller, assuming $k_i = k_{i1}$;

2) when $e_1 < |e(t)| < e_2$, should select PID controller, assuming $k_i = k_{i2}$;

3) when $|e(t)| > = e_2$, delect intergrad unit, select PD controller only, assuming $k_i = 0$.

10.3 控制器设计

经过了卡尔曼滤波后，我们可以开始设计一种基于遗传算法的 PID 控制器。假定 J 为适应度函数，确定 J 的计算公式如下：

$$J = \int_0^\infty (w_1 |e(t)| + w_2 \sup(t) + w_3 u^2(t)) \mathrm{d}t + w_4 t_u + w_5 t_s \qquad (10.11)$$

其中 $e(t)$ 为系统噪声，$u(t)$ 为控制器输出，$\sup(t)$ 为超调量，t_u 为上升时间，t_s 为过渡过程时间，w_1, w_2, w_3, w_4, w_5 为系统权值。根据球杆系统的实际数值，参数可选择如下：

$w_1 = 100$，$w_2 = 1000$，$w_3 = 0.1$，

$w_4 = 200$，$w_5 = 100$。

采用二进制编码，初始种群尺度选 $M = 20$，遗传代数选为 $G = 70$，用轮盘赌选择概率，交叉概率选为 $P_c = 0.7$，变异概率为：

$P_m = 0.03 - [1:1:\mathrm{Size}] * (0.015)/M$。

考虑到实际的球杆系统，可以对一些参数进行优化，参数 $K_p, K_{i1}, K_{i2}, K_d, e_1$，$e_2$ 可以用下面的方法确定：

1）当 $|e(t)| < = e_1$，可以选择 PID 控制器，假定 $k_i = k_{i1}$；

2）当 $e_1 < |e(t)| < e_2$，可以选择 PID 控制器，假定 $k_i = k_{i2}$；

3）当 $|e(t)| > = e_2$，去掉积分选择 PD 控制器，假定 $k_i = 0$。

In GA algorithm, the size of the parameters K_p, K_{i1}, K_{i2}, K_d, e_1, e_2 is selected as following respectively:

[5 10], [6 15], [1 4], [5 10], [0.01 0.02], [0.04 0.06].

So, the experiment curve of ball position is shown as Figure 10.5, the trajectory curve of ball's velocity is shown as Figure 10.6. Simulation and physical control of ball - beam system were mplemented in MATLAB environment. Comparing effects of this algorithm with the conventional PID control, the experimental data proves the superiority of genetic algorithm controller.

在遗传算法中，参数 K_p, K_{i1}, K_{i2}, K_d, e_1, e_2 的尺度分别选择如下：

[5 10]，[6 15]，[1 4]，[5 10]，[0.01 0.02]，[0.04 0.06]。

因此，球的位置的实验曲线如图 10.5 所示，球的速度轨迹曲线如图 10.6 所示。仿真与物理实验都用 MATLAB 语言实现。与传统的 PID 控制器进行比较，可以看出遗传算法与 PID 的结合后实验效果有改进。

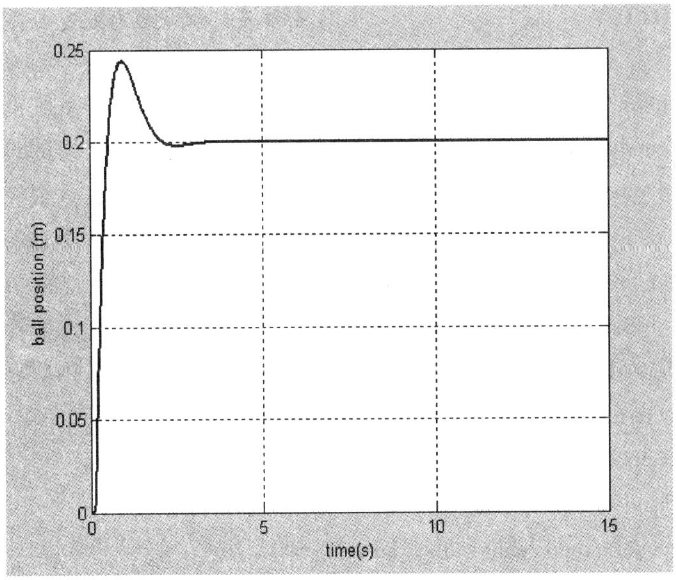

Figure 10.5 Position Response Curve of Ball

图 10.5 球的位置响应曲线

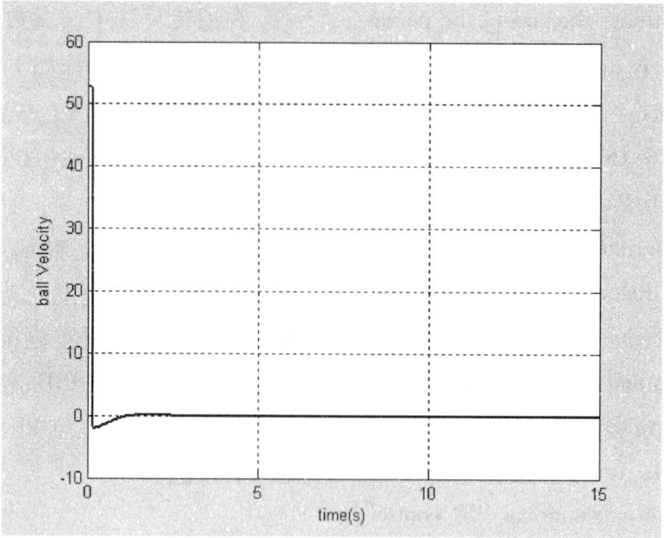

Figure 10. 6 Velocity Response Trajectory of Ball

图 10. 6 球的速度响应曲线

10. 4 Summary

In this Chapter, a kind of GA + PID controller is designed for ball - beam system. The mathematical model for the ball - beam system is established by using Lagrange equation. In order to deal with the noise signal in detection of Ball - beam system, Kalman Filter is adopted and it shows obvious effect in data processing. An integral separation PID + GA control algorithm is designed based on a new kind of fitness function. The experimental data obtained shows the improved control effective.

10. 4 本章小结

本章针对球杆系统的实验，讨论了一种遗传 PID 控制器的设计方法，用拉格朗日法建立了数学模型，用卡尔曼滤波对系统与观测噪声进行了滤除。本章设计了一种积分分离的 PID + GA 控制算法，其适应度函数依据控制系统的指标函数而设定。实验数据显示了算法的有效性。

参考文献

References

[1] Katsuhiko Ogata. Modern Control Engineering [M]. 4thed. New Jersey: Prentice - Hall International, Inc. 2002.

[2] Richard C Dorf, Robert H Bishop. Modern Control Systems (Ninth Edition) [M]. 毕晓普，译. 北京：科学出版社，2002.

[3] 吴麒. 自动控制原理[M]. 北京：清华大学出版社，1996.

[4] 杨位钦，谢锡棋. 自动控制原理[M]. 北京：电子工业出版社，1993.

[5] 胡寿松. 自动控制原理[M]. 4 版. 北京：科学出版社，2001.

[6] 梅晓荣，庄显义. 自动控制原理[M]. 北京：科学出版社，2007.

[7] 韩绍坤，许向阳. 自动控制原理[M]. 北京：北京理工大学出版社，2009.

[8] 姜增如. 自动控制理论实验[M]. 北京：北京理工大学出版社，2010.

[9] 姜增如. 自动控制理论创新实验案例教程[M]. 北京：机械工业出版社，2015.

[10] 戴亚平. 自动控制理论及应用[M]. 北京：中国电力出版社，2015.